Climate Mobility Justice

Elena Giacomelli · Pierluigi Musarò

Climate Mobility Justice

Narratives and Visual Politics of the Panicocene

palgrave
macmillan

Elena Giacomelli
Department of Sociology
and Business Law
University of Bologna
Bologna, Italy

Pierluigi Musarò
Department of Sociology
and Business Law
University of Bologna
Bologna, Italy

ISBN 978-3-032-07906-0 ISBN 978-3-032-07907-7 (eBook)
https://doi.org/10.1007/978-3-032-07907-7

Cover credit: Panicocene. Courtesy of the artist Lara Mezzapelle. Ph. Luigi Russo

This Palgrave Macmillan imprint is published by the registered company Springer Nature Switzerland AG
The registered company address is: Gewerbestrasse 11, 6330 Cham, Switzerland

Foreword

Professor Mimi Sheller, Worcester Polytechnic Institute

As I write this in the summer of 2025 opening the news sometimes feels like reading a dystopian novel. Major wildfires recently scorched Italy, Spain, Portugal, Algeria, Tunisia, and currently are still raging in Canada, Turkey, and across Greece where temperatures are 44C this weekend. In the same news cycle, the Greek prime minister has again suspended asylum applications for migrants from North Africa after a reported 'surge' in numbers. More than 9000 migrants landed in Crete in the last six months, even as 5000 tourists were forced to flee from fires there in July, according to the BBC. These times bring strange conjunctures of climate mobilities, forced deportations, and evacuations.

Catastrophic flash floods are also devastating communities around the world this year, from Afghanistan, Brazil, China and South Asia, to Pakistan, South Korea, Spain, and Texas. Here in the United States where I live these disasters are juxtaposed with the political disaster of the second Trump Administration, which has moved quickly to undo every environmental protection in federal policy, to gut hard-won climate adaptation and carbon mitigation incentives, and to once again push forward the expansion of the fossil fuel industry while hobbling renewable energy. He has a particular disdain for wind power, which disturbs the view from his golf courses. Like some erstwhile Don Quixote, this Ingenious Knight of Mar a Lago tilts at windmills while the world around burns and floods.

At the core of the Make America Great Again (MAGA) blueprint is the breathtaking assault on the rights, protections, dignity, and humanity of people who cross borders. Trump and his Project 2025 acolytes (like Stephen Miller and Russell Vought, assisted by spineless hypocrites like Marco Rubio and J.D. Vance) have unleashed a wave of terror. Gangs of heavily armed and masked Immigration and Customs Enforcement (ICE) agents descend upon small towns, Home Depot parking lots, California farms, Midwestern food processing plants, and even churches, schools, and courthouses across the country (which used to be considered sanctuary spaces) to detain and deport people, with no legal review. Every tenet of domestic and international law has been shredded, while at the very same moment the imposition of new tariffs will disrupt global markets and destroy economies.

Mothers are torn from their children. Grandads and grannies who have lived in the United States for more than thirty years are hauled off to countries they barely remember. Young people who have grown up in the United States are sent to dangerous countries they have never seen. Supposed 'criminal enemy aliens' are deported (many who turn out to have nothing but traffic violations, if that even), many shackled on military flights, with no hearings to foreign prisons in places like El Salvador, Eswatini, Guatemala, South Sudan, and Rwanda, or to

Florida Governor De Santos's wonderful new 'Alligator Alcatraz' in the Big Cypress National Preserve in Ochopee, Florida (unceded territory of the Seminole Indians). Even US citizens and permanent residents are being swept up and locked away. And the National Guard, supported by Marines, were mobilized in Los Angeles against those who protested—many of whom carried stitched together Mexican and American flags to symbolize our deep cross-border connections that run through many families and through Southern California culture at large. While protests continue, and court injunctions proliferate, the ICE terror still grinds on, in the name of the crusade against 'open borders.'

This is why the clear-ringing voices of Elena Giacomelli and Pierluigi Musarò are so inspirational and necessary right now. In the face of burgeoning border walls, violent militarization of borders, and surging ethnonationalism across Europe and America, as well as Asia, they cut through the dominant narratives with a new vision. Dismantling the myths and distortions surrounding climate and migration fanned by populist leaders (who actually represent the interests of capitalist elites not the people), they show us step by step how to build a cosmopolitical movement of ethical planetary action. Drawing on their sharp sociological imagination and deep knowledge of critical approaches and creative interventions, Climate Mobility Justice centers narrative and visual politics as a key to unlocking this political conjuncture to fight for humanity and all earthly creation.

Like the string of events described in my opening paragraphs, usually no clear line of connection is drawn between climate disasters, border policies, and political cataclysms in the mainstream news. Drawing connections devolves to activists, artists, and sociologists like the authors of this book. The headline news barely even mentions that agriculture and fisheries are collapsing in many parts of the world due to extreme heat. That story is left to the environmental reporters or buried in technical science stories. While writers like Bill McKibben and Elizabeth Kolbert make valiant efforts to get these stories out, they seldom connect the climate news to the frontpage headlines about deportations, deaths of migrants at sea, or the rise of ethnonationalist and populist governments.

Drawing on theories of mobility justice and climate justice, Giacomelli and Musarò show us how to arrive at a comprehensive cosmovision of justice that includes not only the interconnectedness of all beings, across every scale, and our responsibility toward future generations. Animating a rich cultural analysis, they help us recognize how dangerous media frames and political discourse based on climate denialism, techno-utopian narratives, and apocalyptic narratives each disarm our ability to act responsibly and ethically. They also show how these dominant narratives lead to three problematic approaches to climate migration: those that frame migrants either as threats, victims, or in need of integration. Building on decolonial and anti-racist analysis by writers such as Andrew Baldwin and Malcom Ferdinand, they help us understand the racialization of the figure of the migrant and its deep colonial origins.

Andrew Baldwin and Malcom Ferdinand, they help us understand the racialization of the figure of the migrant and its deep colonial origins. Building on critiques of climate coloniality, they help think through a transformative practice that expands climate justice from anthropocentric rights to cosmocentric response-abilities. And finally they show us other pluriversal knowledges and ways forward, inspired by Indigenous and decolonial thought, by activist artists and by the everyday diverse acts of creative social participation through which people build bridges, open doors, and create solidarity commons.

Through each chapter they build the case for a new cosmovision that is within our grasp. We just need to open our minds, reframe our reality, and reach out in solidarity with others. While I knew many of the academic and intellectual threads that they are weaving, and in fact have contributed to some of them, I drank in every page like a refreshing draft of cool water: simple, essential, and reenergizing. Their words will touch your heart, chime with your soul, and help hasten the change we need to live through these times and reach a better future. Together—and with care and reverence for multiple embodied experiences both human and more than human, material and spiritual—we can spark a mobile commoning of interdependence that will expand a pluriversal, abundant, and regenerative cosmos like the one we were born into and must hand on to future generations.

I thank Elena and Pierluigi for including me in this conversation and this vision, and invite all of you to join us in making it a reality.

July 2025

Mimi Sheller
Worcester Polytechnic Institute
Worcester, United States

Acknowledgements We are deeply grateful to Nikos Papastergiadis, whose wisdom and imagination inspired this book from its inception beneath the fig tree, and whose companionship guided us toward a more cosmic vision. Our sincere thanks extend to Claudio Boyer for his unwavering availability and clarity, helping to shape our words and thoughts with precision and grace. We are indebted to the collective Reimaging Mobilities, to friends and colleagues—our daily travel companions—whose camaraderie and insights have enriched every step of this journey.

Our profound appreciation goes to Mimi Sheller, whose pioneering work opened the borders of mobility justice for us, and who, during a memorable dinner in Durham, graciously agreed to lend her voice to the foreword. We are also deeply inspired by Andrew Baldwin, whose kindness touched us throughout, and whose brilliant mind advances critical research at the intersections of race, climate change, migration, and humanism. His thoughtful scholarship and insightful afterword have greatly enriched this book.

We wish to honor the artists whose extraordinary visions have illuminated our exploration of climate mobilities. With heartfelt gratitude, we acknowledge Giulia Piermartiri and Edoardo Delille, Anneli Skaar, Lucy + Jorge Orta, Amy Balkin, Michele Lapini, Nick Brandt, Angela Tiatia, Andrea Conte (Andreco), Mali Weil, Jason deCaires Taylor, and Dries Depporter. Your generosity in sharing your remarkable art has allowed us to reimagine the intertwined journeys of people and planet, transforming abstract concepts into vivid, empathetic encounters. Through your work, the unseen becomes visible, and the boundaries of possibility expand.

Finally, this book is dedicated to all family and friends, whose boundless curiosity has encouraged us to embrace questions over answers. And to Jacopo, with the hope that the world you will come to know and shape will not be defined by walls, but by the solidarities that reach across borders.

Competing Interests The authors have no competing interests to declare that are relevant to the content of this manuscript.

Contents

List of Figures

CHAPTER 1

Introduction. Reimagining Narratives at the Intersection of Climate Change and (Im)Mobilities

There's really no such thing as the 'voiceless'.
There are only the deliberately silenced,
or the preferably unheard.
Arundhati Roy (2004)

Words Matter

In 2017, wildlife photographers Paul Nicklen and Cristina Mittermeier traveled to Somerset Island in the Canadian Arctic with the intention of creating images that would capture the full urgency of climate change. During their expedition, they recorded footage of a polar bear ambling across an ice-free landscape and foraging in trash bins. The bear appeared severely malnourished, its coat patchy, and its movements weakened to the point of collapse. The photographers hoped these powerful images would draw global attention to the dangers of climate change. What they did not anticipate, however, was that the footage would be viewed by tens of millions of people worldwide, nor that it would spark intense debate on social media. The main accusation was that the photographers and National Geographic had exaggerated the link between the bear's condition and climate change. In his Instagram post accompanying the video, Nicklen had written, 'This is what starvation looks like,' while referencing scientific concerns that polar bears could face extinction

E. Giacomelli and P. Musarò, *Climate Mobility Justice*,
https://doi.org/10.1007/978-3-032-07907-7_1

within the next century. Though urging viewers to reduce their carbon footprint, he did not claim that climate change was definitively responsible for this particular bear's suffering. However, following National Geographic's release of the video, accompanied by the caption 'This is what climate change looks like,' the video became the most viewed in the publication's history,[1] reaching 2.5 billion viewers globally and prompting intense media coverage and social media engagement.

Some scientists criticized the publication for drawing a direct link between the polar bear's condition and climate change, arguing there was insufficient evidence to attribute the animal's starvation solely to environmental factors—illness or age could also have been factors. In response, National Geographic published an explanatory article by Mittermeier, titled 'Starving-Polar-Bear Photographer Recalls What Went Wrong,' which acknowledged the omission of context. 'Perhaps we made a mistake in not telling the full story,' she wrote, 'that we were looking for a picture that foretold the future and that we didn't know what had happened to this particular polar bear.'[2] The photographers stated that, rather than make hard claims about any individual bear's grim fate, their intent had been to evoke an affective response to the broader risks faced by polar bears: 'We need to get the world talking, and science is obviously not doing that. Science is the foundation, but we need the emotional connection.'

This episode illustrates a persistent tension in climate communication: how to engage audiences emotionally without misrepresenting the facts. In recent decades, scientific research has often outpaced the ability to communicate its urgent message to the broader public. Recognizing this gap, the photographers hoped to elevate climate change to a position of discursive prominence—a place it indeed must reach if meaningful solutions to our global crisis are to be adopted. Their approach to galvanizing public opinion by making the crisis more immediate and emotionally resonant relied on fostering empathetic identification with one of the most iconic and well-travelled symbols of environmental decline.

This appeal to the polar bear raises an enduring ethical question: should we endorse the Machiavellian maxim that 'the end justifies the

[1] https://www.youtube.com/watch?v=_JhaVNJb3ag.

[2] https://www.nationalgeographic.com/magazine/article/explore-through-the-lens-starving-polar-bear-photo.

means,' and legitimize manipulative tactics for a noble cause? Or must we reject these tactics in light of potential reputational harm and unintended consequences?

Documenting the realities of climate change poses significant challenges. The image of a starving polar bear initially appeared to offer a powerful visual metaphor for the future consequences of global warming. However, when National Geographic explicitly linked the bear's starvation to climate change and the image rapidly went viral, the visual association premised on incomplete or potentially inaccurate information ultimately undermined the broader message the photographers sought to convey. Moreover, it may have inadvertently furnished climate change skeptics with further ammunition to challenge both the scientific consensus and the credibility of climate advocates.

This polar bear case emblematizes how powerful imagery can produce outcomes that diverge sharply from the intentions of their creators. It also underscores the limits of relying solely on factual presentation to convey the urgency of climate change. What proves more effective are narratives that strategically use storytelling, visual imagery, and emotional engagement to generate personal resonance. A thoughtful crafting of narrative and visual politics enables audiences to imaginatively inhabit other perspectives, deepening empathy and potentially prompting shifts in attitudes or behaviors. When people emotionally invest in the experiences of others, they re-evaluate their sense of responsibility and care for the collective, often becoming more altruistic, innovative, or pragmatic.

At the same time, as we examine the ways in which scientists, policymakers, activists, and journalists seek to communicate the impacts of global warming, it is essential to recognize that both language and imagery possess the capacity to evoke emotions that may reinforce existing beliefs, regardless of empirical accuracy.

Climate change is one of the most defining challenges of our time. Driven by human activities such as burning fossil fuels, deforestation, and industrial production, it is everywhere all at once, with global temperatures, extreme weather events, and environmental degradation escalating at an alarming rate. Sudden-onset events—typhoons, hurricanes, floods, heat waves—as well as slower-moving phenomena like drought, desertification, and sea-level rise, are triggering large-scale displacement, migration, and (im)mobility, especially in Sub-Saharan Africa, East Asia, the Pacific, and Latin America (Clement et al., 2021). According to the World Meteorological Organization (WMO), over the last fifty years, climate, extreme weather, and water-related events have caused 11,778 disasters and over two million fatalities, with 90% of reported deaths occurring in developing countries. These catastrophes also produced economic losses of approximately US$ 4.3 trillion, with the least developed countries and small island developing states bearing the brunt of the harm relative to the sizes of their economies (WMO, 2023).

The fact that we are living through a climate crisis is substantiated by a robust body of scientific evidence. There is simply no avoiding the issue: summer temperatures flare up, hurricanes wreak havoc up and down the US coast, Europe's rivers burst their banks, and the receding polar ice caps open new shipping lanes. Yet despite the clarity and urgency of the science, the global response has been halting and insufficient. While the facts are widely known, political, economic, and social obstacles—compounded by disinformation—have impeded or delayed decisive action. This disconnect between available knowledge and the feeble measures taken to date highlights a deeper interplay of denial, inertia, and the relative absence of visible short-term consequences, even as the planet's future is put increasingly at risk.

Migration presents a similarly fraught terrain. As we will emphasize throughout this book, migration is both a historical and structural phenomenon, shaped by specific temporal and socioeconomic contexts. It is also profoundly mediated. Both traditional and digital media serve as material and symbolic infrastructures that co-construct how migration is understood and represented globally. Mediatic acts play a pivotal role in shaping public perceptions, actively contributing to the construction of social reality and the development of collective representations. If migration is intricately linked to processes of othering (Hall, 1992; Andreucci & Zografos, 2022) and bordering (Van Houtum & Van Naerssen, 2002; Boas et al., 2024), media practices covering it play a central role in shaping everyday experiences in migratory contact zones together with the broader social imaginaries that precede and shape them. Media logics frequently sensationalize events, framing issues in ways that structure how audiences make sense of the world. These mediated 'maps' are far from neutral; they can either reinforce or challenge prevailing public perceptions regarding migrants' behaviors, intentions, and their perceived impact on host societies (Musarò & Parmiggiani, 2022). This dynamic is particularly pronounced in representations of asylum seekers—humans arriving without prior authorization—but it also extends to other forms of irregular migration and refugee movements (Chouliaraki, & Georgiou, 2022).

Although migration has always been part and parcel of human history, it has gained acute political salience in recent decades. It is now a defining dynamic of the current global order, influencing government policies and shaping public perceptions over the question of who has the right to cross borders. As with climate change, discussions about migration are often clouded by dis- and misinformation, gaps in understanding, and structural bias. In Europe and other so-called Global North countries, politicians and the media tend to focus on irregularized migration,[3] security, and crime, leading to a warped view of the issue. Such distortions lead the public to significantly overestimate the scale of migration and thus to internalize fear-based narratives.

Since 2011, following the so-called North African Emergency, migration to Europe has been increasingly associated with terms such as 'crisis' or 'emergency' (Fontanari & Ambrosini, 2018). This type of framing casts migration not as a structural phenomenon but as an exceptional disruption. In recent decades, narratives, discursive practices, and strategies of representation describing migrants crossing borders as a widespread 'emergency' have proliferated, along with the presupposition that this a social, cultural, and political 'crisis' is best managed at the national and European levels. Seen like this, the phenomenon is brutally simplified and works to justify the advance of restrictive policies, externalized borders, and punitive deterrence strategies (Dastyari et al., 2023).

Two dominant narrative strands prevail: (1) the threat discourse, which constructs immigration as a risk to societal security and stability, and (2) the humanitarian discourse, which highlights the severe hardships and

[3] We tend to prefer the term irregularized migration over irregular migration because it emphasizes the systemic and structural factors that contribute to migrants' irregular status, rather than framing the issue solely around individual actions or choices. The term irregularized highlights how policies, legal frameworks, and enforcement mechanisms actively produce irregularity by excluding migrants from regular pathways or failing to provide adequate protection. This perspective shifts the focus from blaming migrants for their irregular status to examining how institutional practices and governance failures create conditions of irregularity. Moreover, irregularized migration acknowledges the continuum of compulsion that drives migration, recognizing that many individuals are forced into irregular status due to circumstances beyond their control, such as lack of access to visas, restrictive asylum systems, or barriers to regularization. It also reflects the precarious social and economic conditions faced by migrants who are excluded from formal systems of labor, housing, and healthcare.

suffering of migrants. Both narratives were especially pronounced during the 2015–16 influx of asylum seekers and migrants—2.5 million in total—into the European Union. A decade later, these discourses continue to shape public perceptions of asylum seekers and migrants—whether they are traversing the Mediterranean, the US–Mexico border, the EU–Belarus frontier, or the English Channel—and continue to inform the visual representations of forced and irregularized migration across diverse national contexts.

By definition, emergencies are unexpected, short-term, and temporary. On the other hand, climate change and migration are long-term phenomena that require in-depth intersectional analyses and sustained political responses—especially ones that decolonize strategies found to be contributing to the problem. In a context still marked by climate denialism and xenophobia, the interconnected imaginaries of climate change and migration are increasingly fraught with tension. Although these phenomena are long-term structural and political events, in contemporary discourse they are often narrated as circumscribed short-term crises. This has enabled political and media manipulation, distorted perceptions of reality, the proliferation of fake news, and, consequently, the generation of the 'Panicocene' (Giacomelli, 2023; Giacomelli & Cappi, 2025), an era of panic in which the two defining features of contemporary society (the climate crisis and migration) converge within a uniquely distorted narrative—an emergency squared—that generates heightened anxieties and social exclusions, together with the real risks deriving from such narratives (Boas et al., 2024; Baldwin, 2022; Durand-Delacre, 2022). The Panicocene, or 'age of panic,' is not intended to replace frameworks like the Anthropocene or any of its close affiliates (such as the Capitalocene, Plantationocene, Chthulucene, or Wasteocene). Rather, it draws attention to the narrative and representational infrastructures shaping collective social imaginaries in settings of ongoing climate change and shifting patterns of human mobility. Recognizing the processes of framing as crucial elements in social and political transformation (or inertia) requires examining how narratives may either facilitate or hinder (in)justices embedded in processes of bordering and othering mobile populations while delineating future societies as either hostile or welcoming spaces (Fig. 1.1).

Fig. 1.1 Atlas of the New World is a photographic project that visually explores future landscapes shaped by the climate crisis. Using an entirely analogue method, images of these future vistas are briefly projected onto real scenes while photographing unaware subjects, creating a blend of present and future in each photo. This technique metaphorically captures the slipping grip we have on the future. Inspired by Amitav Ghosh's idea of "a crisis of the imagination" in confronting climate change, the project transforms abstract future scenarios into tangible images, highlighting the ongoing flux of time and framing imagining the future as an act of both resistance and hope
In the picture: The Maldives archipelago holds the record for the lowest country in the world. With an average height of only half a meter, it will be the first state to be submerged due to rising sea levels. Courtesy of the photographers Giulia Piermartiri and Edoardo Delille

Reframing the Intersections of Climate and Mobility Justice

The unfolding climate crisis and the intensifying movements of people across and within borders are two of the most urgent and visible phenomena of our time—yet they are often framed as disconnected, exceptional, or even antagonistic. This book brings together the fields of climate justice and mobility justice to critically examine their entanglements and to challenge the dominant narratives that have shaped public discourse, policy, and research in recent decades. Despite their increasing visibility, neither concept has a universally agreed-upon definition.

Climate justice, for instance, has evolved through multiple genealogies: as a legal and ethical discourse, a grassroots movement, and a framework for systemic critique. The phrase is often attributed to Edith Brown Weiss (1989), while its earliest widely recognized use in activist circles can be traced to *Greenhouse Gangsters* vs. *Climate Justice* (Bruno et al., 1999), a report drafted by the US environmental justice movement. These early articulations framed climate justice as an inherently political demand—one that targets the root causes of climate change, including corporate power, fossil fuel dependency, and global inequities produced by colonial capitalism. Since then, the concept has been expanded to include multiple dimensions—distributive, procedural, recognitional, restorative, and intergenerational justice (Sultana, 2022)—and has become central to both academic debates and popular mobilizations.

The concept of mobility justice, developed by sociologist Mimi Sheller (2018), addresses how power and inequality shape the governance and control of movement across multiple scales—from the body and city to the nation and planet. In Shellers' analysis, mobility justice involves multi-scalar entanglements: bodies with unequal mobility (sexed, raced, gendered, classed, abled, aged, etc.); transport system (automobility, cycling, transit, micromobility, aeromobility); urban infrastructures (streets, energy, communication, water, food system); transnational mobility (border regimes, migration, asylum, deportation); and planetary mobility (pollution, waste, CO2; emissions, climate loss & damage). Emerging from both academic research and activist efforts, mobility justice critiques uneven access to mobility and calls for equitable freedom of movement for all, and especially for marginalized communities affected by systemic violence and exclusion. Sheller's concept of 'mobility justice' reflects the following interconnected strands: the right to mobility, the right to live in a healthy environment, and the unequal access to such rights across the globe. In this sense, mobility and climate injustices represent two faces of the same problem.

Yet within mainstream discourse the language of justice has also been systematically co-opted, diluted, or depoliticized. This is especially evident in how media, political rhetoric, and public policy portray climate change and migration. Rather than illuminate the structural causes of environmental degradation and forced displacement, dominant narratives often focus on the symptoms—extreme weather events, displaced populations, rising sea levels—while leaving the economic, political, and social systems that produce these crises unexamined. The public imagination is saturated with apocalyptic images of starving polar bears and sensationalized warnings of mass migration or 'invasion' by so-called climate refugees. What goes missing, of course, are the visual and conceptual links between these outcomes and the predatory logics of industrial development, fossil-fueled capitalism, and global hierarchies built on anthropocentric individualism. This representational imbalance reinforces a deeply uneven geography of visibility and voice, where the causes of the crisis are obscured, and the people most affected are either rendered invisible or framed as threats. The conceptual diversity surrounding climate and mobility justice can produce confusion, but it also reveals a political field of contestation—one where words, meanings, and data must be reclaimed and repoliticized to resist paralysis, fear, and the alienation engendered by alarmist narratives.

This book argues that climate and mobility justice are not only interrelated but co- constitutive—and that their integration is essential for imagining transformative responses to the intertwined crises of the climate crisis and human (im)mobilities. Historically, the two frameworks have developed in parallel but within largely separate spheres: climate justice focusing on ecological harms and their distribution, mobility justice on the governance of human movement and its inequities. Their separation reflects the siloed nature of international governance, as well as those academic disciplines that privilege nation-state perspectives and fixed territorial imaginaries. By bridging these fields, we seek to expose and critique the broader structures of inequality, bordering, and exclusion that shape both climate vulnerability and (im)mobility. We draw on the new mobilities paradigm (Sheller & Urry, 2006) and Sheller's (2018) concept of *mobility assemblages*—constellations of actors, actions, infrastructures, and meanings governed by mobility regimes that dictate who and what may move, where, how, and under what conditions. This lens helps us understand mobility as a mix of freedom and coercion deeply entangled with the histories of empire, racial capitalism, tourism, and labor migration. Moreover, by engaging mobilities theory, we can work through this approach's own blind spots when it comes to climate impacts on existing mobility patterns while continuing to stress the dynamic, two-way relationship between mobility practices and climate change (Baldwin et al., 2019). In reflecting on climate mobility justice, we join other authors in advocating a conceptual shift from 'climate migration' to 'climate (im)mobilities' (Boas et al., 2018), which offers a more nuanced analytical frame encompassing the climate-related practices of communities and individuals and plotting them along a continuum of forced displacement, community relocation, voluntary mobility, and immobilization.

We also aim to disrupt the hegemonic perspective of the Global North in shaping the parameters of both crises. The current political conjuncture is steeped in climate coloniality—the desperate resistance of fuel capital and extractive industries to move toward a just green transition—combined with the militarization of borders and expansion of securitization and surveillance. Migration must not be regarded solely as the symptom of systemic breakdown; it is a constitutive feature of the human experience and an interpretative lens for understanding life and mobility in an increasingly volatile global context. To embrace mobility justice, we must fundamentally shift our perspective—centering the voices and experiences of those most affected by climate change and displacement, beholding them not as victims or invaders, but as subjects of rights, agents of knowledge, and co-creators of alternative futures. Several chapters in this volume draw on Roger Silverstone's (2009) concept of 'media hospitality,' which foregrounds the ethical obligation not only to provide space for the 'other' to speak, but also to listen in ways that affirms their subjectivity, political agency, and transformative potential. To that end, we must evaluate the power of narratives as bordering technologies that determine whose suffering is rendered visible, whose mobility is legitimized, and whose futures are imagined as worth protecting.

To address these challenges, this book adopts a justice-oriented framework that moves beyond conventional distributive or procedural approaches. We foreground the importance of recognitional justice (the right to be acknowledged and represented) and structural justice, which seeks to transform the foundational political and economic arrangements that produce inequality. Our analysis is deliberately multi-scalar—spanning national, international, intergenerational, and cosmovisional dimensions—and attuned to how justice operates differently across uneven landscapes of power. While each of these scales presents opportunities for intervention, all are shaped by geopolitical asymmetries, colonial legacies, and the institutional inertia of global governance. By examining climate-induced (im)mobility through this multi-scalar lens, we uncover both the limits of existing legal and diplomatic frameworks and illuminate the possibilities for resistance, solidarity, and collective action.

How might we reimagine justice across these intersecting terrains of climate mobility? Which kinds of narratives and visual politics are required to construct new imaginaries of climate (im) mobility? Throughout this volume, we aim to answer these questions by supporting alternative mobility imaginaries that build a culture of shared and mobile dwelling places, low-carbon practices and mobile commoning, and respect for human and more-than-human life.

Refusing the Silence, Sharing the Privilege

This book represents not only an academic inquiry but also a political and ethical call to action. We interrogate the alarmist and decontextualized narratives that dominate mainstream discourses on climate change and migration, advocating instead for a reorientation toward stories grounded in care, relationality, and justice. Through critical examination of legal and political frameworks, media representations, activist practices, and artistic expressions, the subsequent chapters elucidate the ways in which diverse actors intend, perceive, depict, contest, reframe, and reimagine climate (im)mobilities. We argue that reclaiming narrative space—through sociological analysis, interdisciplinary collaboration, and creative engagement—is essential for developing more inclusive, equitable, and sustainable responses to the planetary challenges of our time. In so doing, this book encourages readers to engage not only with the crises of our era but also with the epistemological paradigms that have rendered these crises intelligible.

The impulse to write this book emerged from a growing discomfort—and ultimately a refusal—to passively accept how climate change, migration, and their nexus are portrayed in mainstream media and public discourse. What initially compelled us was the profound sense of injustice embedded in the language and imagery that permeate these ubiquitous narratives. Such discourse often denies or distorts historical and scientific realities, effaces the experiences of the most vulnerable populations, and normalizes systemic forms of dehumanization. It obscures the violence enacted through border regimes and policy decisions while severing the intricate, interdependent relationships between humans and non-human beings—including forests, animals, oceans, microbes, and even geological formations. This discursive injustice is not merely rhetorical; it constitutes a foundational component of the ongoing social and ecological crises—a collapse perpetuated to safeguard the comfort and authority of a privileged minority. Acknowledging this dynamic elicits a response that is both intellectual and ethical. Here, we adopt an understanding of responsibility as *response-ability*: the capacity—and the imperative—to think, speak, and act in ways that acknowledge our implication within these systems, particularly for those of us who, by circumstance or design, occupy positions of relative privilege.

We begin with two foundational principles that underpin this book. The first is our own unease, dictated by our situation of privilege. We acknowledge the discomfort—and at times, disorientation—that arises when confronting our own positionality. As white, Western, economically secure, heterosexual researchers living in the so-called Global North, we have not experienced borders as wounds on our own bodies, nor have we endured the existential immediacy of climate collapse in the way many frontline communities already do. For us, the climate struggle often feels projected into the future—a looming horizon rather than a daily reality. This privileged position—rooted in geography, race, and citizenship—raises an urgent ethical question: What do we do with this privilege? Recognition alone is only a first step. Awareness must turn into action. While acknowledging privilege may provoke frustration or a sense of powerlessness, it also provides the conditions for spending it in service of global justice. Our privilege stems, above all, from geography: we were born in Fortress Europe—in what many call the 'right' part of the world. But who gets to choose one's birthplace? To date, (still) no one. The world's social inequities are inscribed on us—embodied inequalities layered across time, populations and histories. We inherit citizenship, the security of a home, and often the shape of our opportunities in life (Vince, 2022). We cannot change the fortune or misfortune of our initial circumstances. But safety and opportunity should not hinge on these arbitrary and unjust inheritances.

Given the social and political nature of research, we believe it is essential to take a stance. As Pierre Bourdieu (2007) reminds us, we cannot remain neutral or detached from the struggles shaping the world's future. In Bauman's (2010) sense, we are responsible subjects: accountable for our words, our knowledge production, and the positions we choose to occupy. Through this work, we strive to deconstruct our own privilege, drawing on intersectional, feminist, and decolonial frameworks to decenter dominant perspectives and contribute to the imagination of more just and inclusive paradigms.

This brings us to the second cornerstone of the book: the imperative to decolonize the imaginary (Dubosc, 2021). In a global context marked by escalating climate anxiety and heightened alarmism around migration, we find it urgent to interrogate the dominant narratives and visual regimes that shape the public understanding of both. Specifically, we seek to uncover what Giuliani (2021) identifies as the *epistemic link* between cultural fantasies of alien invasion—the figure of the unknowable, threatening outsider—and contemporary fears surrounding mass migration to Europe. These fears are not isolated or spontaneous: they are rooted in colonial histories and sustained by the ontologies of the Anthropocene, which privilege control, hierarchy, and extraction. The recurring image of large-scale, uncontrolled, and permanent cross-border movement—what Bettini (2013) describes as the myth of the 'migratory tsunami'—functions as a powerful tool of governance. It frames migration as a crisis, a danger, or a contagion, masking its structural causes and reinforcing securitized, racialized responses. These imaginaries not only depoliticize climate-induced (im)mobilities; they also legitimize border violence, restrict solidarity, and silence alternative understandings of mobility. By situating such narratives within broader historical and political frameworks, this book answers the call to dismantle the epistemologies that uphold injustice. In doing so, we aim to cultivate counter-imaginaries rooted in equity, interdependence, and care—narrative frameworks that resist the logics of fear and domination, and that open space for more inclusive, plural, and relational approaches to climate and mobility justice (Fig. 1.2).

Chapter Outline

This book explores how narratives shape our understanding of the intertwined crises of climate change and human mobility. Each chapter offers a distinct lens on the discursive forces that inform, constrain, or enable just responses to these overlapping global challenges.

Despite the growing engagement of scholars, journalists, and activists with climate-induced human mobility, persistent conceptual ambiguities continue to shape debates on narrative framing and justice advocacy in this arena. The emergent field of climate (im)mobilities research exemplifies considerable interdisciplinary dynamism, integrating insights from

Fig. 1.2 Atlas of the New World. In the picture: California. In California, if greenhouse gas emissions continue to increase, the temperature will rise 8 °C by 2100, the frequency of devastating fires will increase, and 88% of the state will be in extreme drought. Courtesy of the photographers Giulia Piermartiri and Edoardo Delille

environmental science, migration studies, and forced displacement literatures, while simultaneously transcending the limitations of these established disciplinary boundaries. Rather than treating climate (im)mobilities as a simple aggregation of climate change impacts and patterns of (im)mobility, this nascent paradigm conceptualizes them as emergent phenomena in their own right. Drawing on the principle of non-additive complexity—where the synthesis of elements produces a qualitatively distinct ontological category—climate (im)mobilities are understood as unique socio-ecological configurations, characterized by specific causal mechanisms, feedback loops, and governance challenges that cannot be reduced to their constituent parts.

The recent conceptual shift toward framing climate (im)mobilities as emergent socio-ecological systems (Boas et al., 2019) underscores the need for parallel innovations in the construction of justice-oriented narratives. This reconceptualization challenges reductionist causal frameworks and foregrounds the irreducible complexity of lived experience in the context of environmental change. Accordingly, the intersection of climate change and human (im)mobility emerges as one of the most urgent challenges of our era. To confront it, we need a multifaceted, transdisciplinary approach that moves beyond conventional paradigms of justice.

At the heart of our inquiry lies a critical engagement with the role of storytelling—understood broadly as media, policy, and cultural expression—in framing both the causes and consequences of climate-induced (im)mobilities. Chapters 2, 3, and 4 delve into the powerful influence of media narratives, political rhetoric, and cultural imaginaries on the construction of perceptions and the formation of policy approaches. These chapters critically examine the dominant 'crisis' imaginary—a framing that often oscillates between the securitization of borders and humanitarian portrayals of displaced people. As this dual logic of fear and care unfolds, it reveals the inherent tension in contemporary responses to climate (im)mobility (Chandler, 2014; Chouliaraki & Georgiou, 2022). Taken together, these chapters argue for a shift beyond simplistic binaries by embracing a multidimensional perspective rooted in the principles of mobility and climate justice. The final section of the book (Chapters 5 and 6) turns to creative and institutional efforts that challenge or rearticulate dominant discourses. These chapters examine how international organizations, media campaigns, and contemporary art practices either reproduce or contest mainstream narratives. Whether it is by delving into the strategic communication of the International Organization for Migration (IOM) and international NGOs, or through the imaginative work of artists, these chapters highlight how the visual politics of the Panicocene can reproduce stereotypes and toxic narratives that discriminate against vulnerable populations—or, alternatively, open the social imaginary to visions that foreground care, solidarity, and cosmopolitan ethics over securitized responses. By tracing these narrative shifts, this book invites readers to consider how more just and inclusive futures might be imagined—and enacted—at the troubled interface of climate change and human mobility.

Throughout this book, you will encounter selected artworks that serve as visually explanatory representations of the key concepts explored in each chapter. These artistic works are not merely decorative additions but essential visual tools that illuminate complex ideas related to climate and mobility justice, the power of narratives, climate (im)mobilities, and other central aspects of our discussion. By employing decolonial and intersectional frameworks, these visual narratives actively challenge dominant historical and cultural hierarchies, recentering marginalized perspectives and exposing the interconnected systems of power that shape climate, mobility, and social inequities (Smith, 2012). Research demonstrates that visual explanations significantly enhance learning outcomes, particularly when dealing with mechanisms and processes outside our immediate experience (Mayer, 2009). Our brains take in and interpret information through two distinct but complementary channels—one for verbal content, such as words and language, and another for visual content, like images and spatial relationships. When both these channels are activated at the same time, it creates a richer learning experience that helps individuals grasp complex ideas more deeply and remember them more effectively (Paivio, 1990). In our book, images serve the crucial dual purpose of conveying information while simultaneously illustrating it, providing context and making our arguments clearer to audiences (Kress & van Leeuwen, 2021). The inclusion of these visual narratives ensures that overlapping struggles—such as those tied to race, gender, class, and Indigeneity—are not siloed but instead portrayed as interdependent dimensions of climate mobility justice (Crenshaw, 1991). The artworks included here function as powerful platforms for inference, providing checks for completeness and coherence while making different features of justice visible and tangible.

Chapter 2 offers a critical examination of the terrain traversed by climate and mobility justice. Drawing from justice theories and the new mobilities paradigm, it contends that while climate change is often narrated as a universal challenge, its impacts are unevenly distributed, reflecting and reinforcing entrenched global and local inequalities. The chapter dissects how climate-induced disruptions amplify existing social, economic, and political vulnerabilities—disproportionately affecting historically marginalized communities, Indigenous peoples, and those least responsible for ecological degradation. It critiques dominant policy frameworks—both national and international—that emphasize technocratic, market-driven solutions while neglecting the structural roots of injustice. Simultaneously, it interrogates the uneven geographies of mobility, revealing how state-centric border regimes, visa policies, and migration controls reproduce global hierarchies of race, class, and citizenship. By integrating multidimensional understandings of justice—national, international, intergenerational, and cosmovisional—the chapter advances a call for a unified, justice-oriented framework that challenges epistemic and institutional boundaries between climate and migration governance. In doing so, it enacts the kind of multi-scalar, intersectional, and more-than-human approach necessary for rethinking justice in a world marked by environmental precarity and unequal mobility.

Chapter 3 analyzes the discursive architectures through which climate change and migration are rendered intelligible in public discourse. Drawing on sociological and philosophical insights, including Rancière's (2009) notion of the 'distribution of the sensible,' it exposes how dominant media and political narratives selectively frame these phenomena—often through apocalyptic, technocratic, or securitized lenses. The chapter grapples with the cognitive and communicative traps set by denialist, techno-utopian, and apocalyptic narratives. Our analysis shows how these dominant scripts either minimize the crisis or foster fatalism and inaction, revealing how climate narratives become depoliticized when presented as distant or inevitable, thereby perpetuating the evasion of accountability while reinforcing elite interests. In parallel, migration is persistently framed as a crisis, oscillating between humanitarian and securitarian logics that cast migrants as either victims or threats and legitimize exclusionary state responses. These binary constructions, deeply embedded in Western-centric epistemologies, obscure the structural drivers of displacement and deny the agency of mobile populations. The chapter critiques the rise of neoliberal integration discourses that repackage migrants as entrepreneurial assets, masking systemic inequalities under a utilitarian logic. In response, it calls for a justice-centered narrative politics—one that destabilizes symbolic borders, amplifies subaltern voices, and fosters solidaristic imaginaries capable of galvanizing transformative social and policy change.

Chapter 4 addresses the multifaceted dynamics of climate-induced (im)mobility, situating them within broader historical and structural contexts shaped by colonialism, capitalism, and environmental degradation. By engaging with frameworks such as the Anthropocene and Capitalocene, the chapter interrogates the geopolitical and racialized dimensions of mobility and immobility in the context of climate change. It challenges dominant security-centric narratives that portray climate migrants as looming threats, revealing how such framings sustain militarized border regimes and obscure the root causes of displacement. Critically, it underscores that not all affected individuals move; many remain trapped or choose to stay due to cultural, economic, or political constraints—and yet are rendered invisible in policy and media discourse. This chapter's intersectional lense foregrounds feminist, decolonial, and

Indigenous epistemologies that resist universalist assumptions and highlight plural experiences of (im)mobility. It advocates for justice frameworks that extend beyond the right to mobility, to include environmental, cultural, and territorial rights while calling for an ethics of care and solidarity. Here we set up the conceptual groundwork needed to analyze how actors across sectors can reshape public opinion and policy responses through alternative narratives and practices.

Chapter 5 examines the discursive power of information and awareness campaigns in constructing social imaginaries around climate (im)mobility. It analyzes how communication efforts led by states, intergovernmental bodies, and civil society organizations mediate public understandings of the relationship between environmental change and human movement. Drawing on critical discourse and visual analysis, we demonstrate how many campaigns reproduce dominant crisis framings that oscillate between victimization and securitization, thereby reinforcing hierarchies of visibility and legitimacy. In contrast, campaigns grounded in decolonial, feminist, and participatory frameworks challenge these logics by foregrounding migrant agency, systemic critique, and calls for climate reparations. Eight emblematic campaigns are dissected in order to trace how narrative form, aesthetic strategy, and institutional authorship shape their ethical and political effects. It argues that such campaigns are performative acts that delineate moral geographies—determining whose suffering is rendered visible, whose claims are legitimized, and whose futures are deemed worth protecting. The chapter concludes by advocating for a justice-oriented communication paradigm that transcends victim/hero binaries and centers interdependence, historical accountability, and inclusive imaginaries of belonging.

Chapter 6 explores the generative role of contemporary art in contesting dominant framings of climate change and human mobility. It examines how artists engage with themes of displacement, resistance, and ecological precarity to craft alternative imaginaries that transcend the limitations of policy discourse. Focusing on a diverse range of artistic interventions—from performance and installation to participatory and multispecies practices—the chapter argues that art constitutes a critical space for emotional engagement, epistemic disruption, and political imagination. Close readings of works by Angela Tiatia, John Akomfrah, Yinka Shonibare, and others, show how engaged artistic practices disrupt securitarian and humanitarian tropes by reclaiming agency, resilience, and relationality. Problematizing the politics of victimhood (Chouliaraki,

2024), these practices challenge the spatial and temporal logics of Western modernity, reframe migration as adaptation and survival, and amplify Indigenous and subaltern worldviews often erased in global governance frameworks. The chapter situates art not merely as representation but as an intervention in the public sphere and conduit to new solidarities. In doing so, it underscores the potential of art to animate justice-driven responses to climate (im)mobility and inspire cross-sectoral collaboration toward more inclusive planetary futures.

The concluding chapter synthesizes the theoretical insights and empirical cases explored throughout the book to argue for a radical reorientation of how climate and mobility are conceived, governed, and narrated. Drawing together themes of justice, narrative power, and epistemic plurality (Escobar, 2018), it reflects on the limitations of existing paradigms and the transformative potential of alternative frameworks rooted in care, solidarity, and planetary responsibility. The conclusion reaffirms the urgency of dismantling dominant imaginaries that naturalize inequality and exclusion and calls for the cultivation of cross-disciplinary and cross-cultural alliances in pursuit of climate mobility justice. Our greatest hope is that the right to remain in place and the right to move freely across borders may be defended equally. To do this, we will insist on forms of human and more-than-human empathy and relationality animating the cosmovisional 'right to breathe' (Mbembe, 2021).

References

Andreucci, D., & Zografos, C. (2022). Between improvement and sacrifice: Othering and the (bio)political ecology of climate change. *Political Geography, 92*, 102512. https://doi.org/10.1016/j.polgeo.2021.102512

Arundhati, R. (2004). *2004 Sydney peace prize*. Sydney Peace Prize. https://sydneypeacefoundation.org.au/peace-prize-events/lectures/.

Baldwin, A. (2022). *The other of climate change: Racial futurism, migration, humanism*. Rowman & Littlefield.

Bauman, Z. (2010). *Liquid times: Living in an age of uncertainty (Repr)*. Polity Press.

Bettini, G. (2013). Climate barbarians at the gate? A critique of apocalyptic narratives on 'climate refugees'. *Geoforum, 45*, 63–72. https://doi.org/10.1016/j.geoforum.2012.09.009

Boas, I., Farbotko, C., Adams, H., Sterly, H., Bush, S., van der Geest, K., Wiegel, H., Ashraf, H., Baldwin, A., Bettini, G., Blondin, S., de Bruijn, M., Durand-Delacre, D., Fröhlich, C., Gioli, G., Guaita, L., Hut, E., Jarawura, F. X.,

Lamers, M., et al. (2019). Climate migration myths. *Nature Climate Change, 9*(12), 901–903. https://doi.org/10.1038/s41558-019-0633-3
Boas, I., Farbotko, C., & Bukari, K. N. (2024). The bordering and rebordering of climate mobilities: Towards a plurality of relations. *Mobilities, 19*(3), 521–536. https://doi.org/10.1080/17450101.2023.2279095
Bourdieu, P. (2007). *La misère du monde*. Éditions du Seuil.
Brown Weiss, E. (1989). *In fairness to future generations: International law, common patrimony and intergenerational equity: Vol. yearbook of international environmental law*. United Nations university Transnational publ.
Bruno, K., Karliner, J., & Brotsky, C. (1999). *Greenhouse gangsters vs. climate justice*. Transnational Resource and Action Center.
Chandler, D. (2014). *Resilience: The governance of complexity*. Routledge. https://doi.org/10.4324/9781315773810
Chouliaraki, L. (2024). *Wronged: The weaponization of victimhood*. Columbia University Press.
Chouliaraki, L., & Georgiou, M. (2022). *The digital border: Migration, technology, power*. New York University Press.
Clement, V., Rigaud, K. K., de Sherbinin, A., Jones, B., Adamo, S., Schewe, J., Sadiq, N., & Shabahat, E. (2021). Groundswell part 2: Acting on internal climate migration. World Bank. https://openknowledge.worldbank.org/handle/10986/36248.
Crenshaw, K. (1991). Mapping the margins: Intersectionality, identity politics, and violence against women of color. *Stanford Law Review, 43*(6), 1241. https://doi.org/10.2307/1229039
Dastyari, A., Nethery, A., & Hirsch, A. L. (Eds.). (2023). *Refugee externalistation policies: Responsibility, legitimacy and accountability*. Routledge, Taylor & Francis Group. https://doi.org/10.4324/9781003167273
Dubosc, F. O. (2021). *Decolonizzare l'immaginario Decolonizzare il tempo*. Clinica Della Crisi. https://clinicadellacrisi.home.blog/2021/02/15/decolonizzare-limmaginario-decolonizzare-il-tempo/.
Durand-Delacre, D. (2022). (Mis)representing climate mobilities: Lessons from documentary filmmaking. *Journal of Ethnic and Migration Studies, 48*(14), 3397–3415. https://doi.org/10.1080/1369183X.2022.2066260
Escobar, A. (2018). *Designs for the pluriverse: Radical interdependence, autonomy, and the making of worlds*. Duke University press.
Fontanari, E., & Ambrosini, M. (2018). Into the interstices: Everyday practices of refugees and their supporters in Europe's migration 'crisis'. *Sociology, 52*(3), 587–603. https://doi.org/10.1177/0038038518759458
Giuliani, G. (2021). *Monsters, catastrophes and the anthropocene: A postcolonial critique*. Routledge.
Hall, S. (1992). The west and the rest: Discourse and power. In S. Hall & B. Gieben (Eds.), *Formations of modernity* (pp. 275–331). Polity Press.

Kress, G. R., & Leeuwen, T. van. (2021). Reading images: The grammar of visual design (Third edition). Routledge Taylor & Francis Group.
Mayer, R. E. (2009). *Multimedia learning* (2nd ed.). Cambridge University Press. https://doi.org/10.1017/CBO9780511811678
Mbembe, A. (2021). The universal right to breathe. *Critical Inquiry, 47*(2), 58–62.
Musarò, P., & Parmiggiani, P. (2022). *Ospitalità mediatica. Le migrazioni nel discorso pubblico.* FRANCOANGELI.
Paivio, A. (1990). *Mental representations.* Oxford University Press. https://doi.org/10.1093/acprof:oso/9780195066661.001.0001
Rancière, J. (2009). *The politics of aesthetics: The distribution of the sensible (reprinted).* Continuum.
Sheller, M. (2018). *Mobility justice: The politics of movement in the age of extremes.* Verso.
Sheller, M., & Urry, J. (2006). The new Mobilities paradigm. *Environment and Planning A: Economy and Space, 38*(2), 207–226. https://doi.org/10.1068/a37268
Silverstone, R. (2009). *Mediapolis: La responsabilità Dei media nella civiltà globale.* V & P.
Smith, L. T. (2012). *Decolonizing methodologies: Research and indigenous peoples (second edition).* Zed.
Sultana, F. (2022). The unbearable heaviness of climate coloniality. *Political Geography, 99*, 102638. https://doi.org/10.1016/j.polgeo.2022.102638
Van Houtum, H., & Van Naerssen, T. (2002). Bordering, ordering and othering. *Tijdschrift voor Economische en Sociale Geografie, 93*(2), 125–136. https://doi.org/10.1111/1467-9663.00189
Vince, G. (2022). *Nomad century: How climate migration will reshape our world (first U.S. edition).* Flatiron Books.
WMO. (2023). *State of the global climate 2023.* World Meteorological Organization. https://wmo.int/publication-series/state-of-global-climate-2023

CHAPTER 2

The Legal and Political Challenges of Climate and Mobility Justice

Introduction

In this chapter, we will insist that while climate change affects everyone, it does so unevenly. This disparity speaks to the profound effects of the crisis on human rights and social justice, as well as the dire need for climate justice, which seeks to address the economic, social, legal, and political dimensions of climate change. Among these demands, climate justice emphasizes holding major polluters accountable, ensuring equitable access to resources for adaptation and mitigation, as well as including and amplifying the voices of marginalized groups in climate decision-making.

We will explore how the concept of climate justice is understood in a multiplicity of ways, each functioning as sides of a mirror reflecting different visions of justice. Climate change is a multifaceted issue that not only reflects and perpetuates structural injustices but also poses a challenge to understanding, given that it is perceived differently by individuals based on their unique perspectives.

At the very same time, and as we will see in greater detail throughout the book, though migration in general has long been perceived as a problem, due to the escalating climate crisis it is now emerging as a pivotal issue within the framework of climate justice discourse. Sudden and slow-onset events are directly impacting human (im)mobility. Contemporary climate displacement is reaching unprecedented levels, with annual

E. Giacomelli and P. Musarò, *Climate Mobility Justice*,
https://doi.org/10.1007/978-3-032-07907-7_2

increases in the global number of people who are forced to move (UNHCR, 2022). According to the Internal Displacement Monitoring Centre (IDMC), in 2022 there were over 32 million new displacements due to disasters, the vast majority of which were related to climate and environmental factors with between 3.3 and 3.6 billion people already living in areas considered highly vulnerable to climate change (IDMC, 2022).

Despite constant references to 'climate refugees,' this category remains formally unrecognized. Moreover, any discussion of migration that fails to address the processes of 'othering'and 'bordering'is fundamentally flawed (Chouliaraki & Georgiou, 2022). This oversight highlights how border control systems perpetuate vulnerability to violence and systemic exclusion among certain populations, and underscores freedom of movement, or lack thereof, as a key indicator of social stratification in our globalized world and a central concern within the framework of mobility justice. As Zygmunt Bauman (2013) argues, mobility is no longer a universal right, but a privilege unevenly distributed across social and economic hierarchies. The rich and powerful enjoy unrestricted movement, symbolizing their global citizenship and access to opportunities. In contrast, the marginalized are often immobilized by borders, surveillance, restrictive migration policies, and confined to precarious spaces. This unequal distribution of mobility reinforces existing inequalities, with freedom of movement becoming a tool for maintaining and legitimizing social stratification in a world increasingly shaped by the dynamics of globalization and neoliberalism (Wihtol de Wenden, 2015).

Stressing how climate justice and mobility justice have historically been siloed due to their distinct focus areas, we will argue in the last part of this chapter, and more broadly in the third chapter, that addressing these interconnected challenges requires a unified approach. Exploring climate justice and mobility justice together allows us to uncover deeper layers of inequality while also designing alternative perspectives that could benefit both humans and the planet. Thus, rather than beginning our analysis from the sedentary perspective of nation-states and societies, or even that of individuals and groups, we will focus on the organization of power around systems of governing mobility, immobility, flows, and borders at various scales.

We will use different theories of justice to outline different approaches, and to show how each method can reveal different dimensions of the issue—-procedural, recognitional, corrective and/or restorative, distributive, and intergenerational (Sultana, 2022a). While these dimensions are often interconnected and may overlap in practice, they offer complementary perspectives that together provide a more thorough and well-rounded understanding of climate justice in our present conjuncture.

To extend the concept of justice to the context of climate change and migration, while also integrating the vast literature and intense contemporary debate on the topic, we will explore the different approaches across scales, including national, international, intergenerational, and cosmovisional. At each level, we can find specific tools and policies for responding to both the causes and effects of climate change and migration (legal frameworks, global agreements, economic mechanisms of adaptation and compensation, cultural practices, etc.). To be sure, all these approaches face limitations when dealing with thoroughly global and multidimensional phenomena. Ultimately, different actors' visions and decisions reflect the tensions between differing national interests, economic disparities, and historical and colonial power dynamics.

In attending to the current expectations surrounding climate-induced (im)mobility, our multi-scalar model sheds light on the limitations of existing policies as well as the historical and conceptual gaps that underpin popular proposals for how to address climate and mobility justice today. Climate and mobility justice usually intersect within global agreements, international humanitarian law, and diplomatic negotiations. However, the paradigms behind these policies contain elements that are often overlooked or obscured in public discourse.

By highlighting how the current debate mainly foregrounds the recognition of climate refugees and the establishment of global protocols for their rights and resettlement (like the 1951 Refugee Convention), while advancing market-driven solutions (like the *Loss and Damage* tool or the *Green Climate Fund*) to help vulnerable countries cope with the effects of climate change and prevent forced migration, this chapter presents a critique of paradigms that treat displaced people as passive victims or that frame economic growth as limitless and as an end in itself.

In some cases, these well-intentioned measures even seek to prevent people impacted by the climate crisis from relocating and seeking refuge, frequently by building walls and militarizing borders. This chapter will argue that these visions of justice are too narrow and simplistic, motivated more by fear and the protection of the wealthy and their privileges, rather than by principles of equality, dignity, and fairness. As such, these high-profile approaches align more closely with ideas of security and control than those of equity and justice.

Taking a broader perspective requires a multi-scalar approach that integrates national action, international cooperation, long-term thinking, and a more-than-human cosmovision. By combining these practices, we can ensure that justice is not only about the rights of those displaced or affected today but also considers the future generations and the environment itself, thus creating a more holistic, fair, and sustainable framework for addressing the challenges of climate change and its attendant migratory consequences.

Finally, by applying the new mobilities paradigm (Sheller & Urry, 2006) to the domain of climate mobilities, this study will critically examine how climate mobility regimes address and protect the rights to dwell and move of populations disproportionately affected by climate change, particularly those who have contributed minimally to its causation. The concept of climate mobility regimes will help us illuminate the multiplicity and intersectionality of climate change-related human mobility, which also involves immobility, relocation, circular mobility, etc. This broader concept encompasses the full spectrum of human movement in the context of climate change, highlighting differences of distance, duration, and degree of voluntariness, even in cases where people are unable to leave or choose to stay where they are.

What Do We Talk About When We Talk About Climate Justice

Climate change presents, above all, a moral challenge that, directly or indirectly, is not sufficiently acknowledged in most policy discussions. Climate change is often treated as a mere technical problem, either approached as a problem of public administration to be addressed by policymakers, or as a matter of economic policy better left to the economists. In these contexts, it is rarely mentioned that the climate crisis has specific historical causes leading to differentiated responsibilities. The populations—and ecosystems—that suffer the most from the consequences of climate disruption are those that have contributed the least to disturbing the balance of the planet. These groups are already more vulnerable 'due to factors such as interconnected socio-economic development models, unsustainable use of oceans and land, social inequalities, and forms of marginalization, as well as historical and ongoing dynamics like colonialism and resource governance' (IPCC, 2022, p.17). Thus, on the one hand, environmental challenges exacerbate already existing economic and social inequalities that might have their root causes elsewhere. On the other hand, we need to consider that injustice in settings of climate change means that those least responsible for greenhouse gas emissions are the ones who are made most vulnerable to the consequences of climate change while possessing the fewest resources to adapt. For this reason, climate change must first and foremost be considered a moral issue to be reframed, as activists and scholars have done for more than two decades, in terms of in/justice.

Let us try to briefly trace the history of this debate to better understand its implications and limitations. In the public arena, climate change historically finds its initial place within the broader 'environmental question,' a discourse that became more defined after the Second World War, when environmental threats tied to economic expansion led those who feared for their lives and well-being to unite and act in unison (Pellizzoni, 2007). In the 1970s, a series of publications, including the *Limits to Growth* report conducted by the Massachusetts Institute of Technology on behalf of the Club of Rome, sparked a debate on the 'dilemmas of humanity,' out of which emerged the need for a 'consciousness that urgently demands radically new visions and approaches to addressing the complex, bewildering, and unprecedented issues facing all of human society' (Meadows and with Club de Rome 1979, p. 12). There was also a call to 'expand the horizon of our interests both temporally and spatially, shifting them away from immediate or local issues [...] to consider those that are far more complex and important, concerning the organization of life for four, five, six, or seven billion inhabitants on our planet under reasonable conditions of well-being, justice, and harmony with Nature' (Meadows and with Club de Rome 1979, p. 13).

The emergence and delineation of these ethical issues, accompanied by an increasingly apocalyptic tone surrounding climate change and its socially constructed representation, gave rise to new forms of civic responsibility and subjectivity. Moreover, with the spread of new social movements (pacifist, student, feminist, hippie) in the late 1960s, the environmentalist protests, while still mostly manifesting at a somewhat situated, local scale, gradually turned into a mass movement (McNeill & Engelke, 2016, pp. 173-174).

Governmental and intergovernmental engagement with environmental issues began to intensify, accompanied by a proliferation of international organizations, conferences, and summits pronouncing agreements, many of which would later become routine references. In 1988, 1 year after the adoption of the Montreal Protocol, wherein signatory parties committed to limiting the production of substances harmful to the ozone layer, the IPCC (Intergovernmental Panel on Climate Change) was established. This marked the United Nations' effort to provide governments worldwide with a comprehensive scientific overview of the current state of knowledge on climate change.

A major turning point came with the 1992 Earth Summit in Rio de Janeiro, which brought together representatives from 172 governments (including 108 heads of state) and 2400 representatives from nongovernmental organizations. A parallel Forum organized by NGOs attracted an additional 17,000 participants. In terms of media impact, it was an unprecedented event—one that catalyzed a shift toward transnational networking among local environmental and ecological movements, which increasingly came to play a central role in contesting the policies and discourses of government elites.

The Rio Conference also marked the emergence of a new framework for climate governance, structured around the annual Conferences of Parties (COP), which bring together the signatory countries of the United Nations Framework Convention on Climate Change drafted and adopted during the Earth Summit.

Although several theorists tend to regard climate justice as a specific branch or application of climate ethics, the term itself—often associated with the question of how to assign responsibilities through an international treaty framework (Moellendorf, 2014)—began to gain traction in the late 1990s. This shift followed a wave of activism by social and environmental justice movements that emerged in response to the operations of the fossil fuel industry and, later, to what many participants perceived as the failure of global climate governance—particularly evident at the COP15 in Copenhagen in 2009 (Schlosberg & Collins, 2014).

Over the last decade, however, the field has seen rapid growth, largely driven by the persistent failure to address the stratifying effects of climate challenges. These include increasingly limited and unequal access to food, water, and energy. Such pressing privations call for new approaches and methods capable of confronting the systemic inequities of climate change and its impacts. While understandings of climate justice—and its practical implementation—vary greatly, most scholars agree that its core emphasis on fairness and equity offers a robust framework for navigating the uneven landscape of climate vulnerability and responsibility.

But what exactly is climate justice? The concept has generated a substantial body of literature, though one marked by a certain degree of conceptual uncertainty. Despite ongoing ambiguity, certain common denominators can be discerned. Climate justice, first and foremost, stipulates a relational understanding of justice—namely, the recognition that justice needs to be actualized *between* different actors. Therefore, it becomes essential to identify both the recipients and the bearers of justice while establishing which entities—whether individuals, social groups, states, or even non-human actors—can assert claims against others. As Beckman and Page show in their *Perspectives on Justice, Democracy, and Global Climate Change* (2008), the literature on climate justice is organized around several defining features, including the identification of recipients, burden-bearers, attention to temporal aspects, and the articulation of normative claims. On this basis, several authors distinguish between the principal dimensions of climate justice, commonly identified as procedural, recognitional, corrective and/or restorative, distributive, and intergenerational (Sultana, Sultana 2022b).

Let us look at each in turn. First, the concept of procedural justice is rooted in the principles of fairness articulated by the philosopher John Rawls in *A Theory of Justice* (2005) and later by the economist Amartya Sen in *The Idea of Justice* (2009), where a strong emphasis is placed on inclusive and equitable decision-making processes. Procedural justice has been defined as 'justice in the way outcomes are brought about, including who participates and is heard in the processes of decision-making' (Damon Matthews et al., 2021, p. 2913). Participation in the decision-making processes can take place across various scales: from international negotiations between states, to participation in the formulation of national climate strategies, to the requirement of prior and informed consent in the implementation of carbon market project areas throughout different regions of the world.

Second, recognitional justice, which is inherently linked to procedural justice, deals with the question of who may be recognized as a legitimate actor with the right to participate in decision-making (Fraser, 2012). This mode of justice emphasizes the importance of acknowledging the dignity, rights, and identities of individuals who have suffered from historical wrongdoing. Although it does not necessarily demand material compensation for past wrongs, it asserts that the acknowledgment of these injustices constitutes a vital first step toward fostering more equitable futures. For instance, Indigenous communities in the United States invoke recognitional justice for the forced removal from ancestral lands and other forms of dispossession (Whyte, 2016). Similarly, communities affected by environmental displacement have sought recognition of their spiritual and cultural ties to sacred ancestral sites, as well as the incorporation of their knowledge systems and core values—such as trust, reciprocity, responsibility, and shared identity—into the policy frameworks that impact them.

Third, restorative (or corrective) justice addresses the rectification of historical injustices that have contributed to current climate-related harms. It seeks to establish accountability for present-day losses and damages and to ensure appropriate forms of redress and compensation (Robinson & Carlson 2021). This dimension is reflected in the UNFCCC's 'Polluter Pays Principle' and underpins discussions around climate reparations in the ongoing climate policy negotiations around *Loss and Damage* (Jackson et al., 2023).

Fourth, distributive justice refers to the unequal ways in which the impacts of climate change are perceived and experienced, both between and within societies—disparities that are often shaped by intersecting dynamics of race, gender, class, and species (Newell, 2005). Intergenerational justice, by contrast, concerns the extent to which present decisions consider climate-related consequences on future generations.

While references are often made to Rawlsian liberal traditions of justice, which presume that recognition and participation are absorbed within the distributive justice paradigm, in practice, these dimensions are deeply interconnected. Recognition is a prerequisite for meaningful participation and representation in climate politics and thus for procedural justice as well. Furthermore, questions arise regarding which knowledge systems should inform decision-making—a reflection of entrenched hierarchies of recognition embedded in knowledge production and unequal access to centers of power. In this context, the importance of cognitive justice has also been brought to the fore (Visvanathan, 2007) by calling for respect for multidimensional ways of understanding the relationship between climate and society (Escobar, 2018). This includes, most urgently, Indigenous cosmologies that promote notions of 'justice for nature' by challenging dichotomies between humans and nature (Bhambra & Newell, 2023).

Climate change serves as a paradigmatic example of distributive injustice, as the Global North—responsible for the majority of historical greenhouse gas emissions—has disproportionately benefited from fossil-fueled economic development. Meanwhile, those who are disproportionately harmed by climate change are the states and communities who have contributed least to the problem. As González (2017, p. 113) argues, 'Climate change raises issues of procedural injustice because the North dominates the institutions of global economic and environmental governance and frequently ignores Southern perspectives and priorities. Climate change emblematizes corrective injustice because those who bear the greatest climate change-induced harm (including indigenous peoples and small island states) have been unable to obtain compensation or redress for their wounds. Finally, climate change is inextricably linked to broader social injustice, including an economic order that exacerbates poverty and inequality while exceeding the limits of the planet's finite ecosystems.'

Moreover, the concept of justice (or injustice) is articulated differently across various social and institutional domains—including by social movements, NGOs, academics, and policymakers (Schlosberg & Collins, 2014). Current approaches to climate justice thus encompass a broad and evolving set of commitments: from the imperative to dismantle and alleviate the unequal burdens imposed by climate change, to the obligation to address the disproportionate burden of the climate crisis on poor, the Indigenous, and marginalized communities; from the recognition that the most vulnerable are often the most deserving of protection and support, to the demand for dismantling the entrenched power of the fossil fuel industry and other extractive structures.

With this definitional fluidity and plurality in mind, the following sections aim to illuminate the current debates about climate justice by exploring multiple scalar perspectives: national, international, intergenerational, and cosmovisional. These scales are not mutually exclusive; rather, they represent complementary perspectives that collectively contribute to a more holistic understanding of climate justice in the contemporary world.

Intra-Societal Justice: The Role of Intersectional Factors at the National Level

Several activists and scholars have drawn attention to the problem of intra-societal injustice, which refers to disparities among social groups based on age, class, gender, race, disability, and other identity markers. Harris et al. (2013, pp. 294-300), among others, argue that we must 'open up the traditionally closed box of "the state", [to] see that the real divide is not so much between developed and developing states as it is between affluent and poor people.' Intra-societal injustice captures how different groups within and across societies experience unequal exposure to the impacts of climate change—often despite having contributed minimally to its genesis.

More generally, activists denounce the built-in growth imperative and increasing concentrations of wealth intrinsic to modern capitalism as the primary drivers of both social and environmental crises. They call for a transition to a more inclusive and democratic economic system as a precondition for meaningfully addressing the climate emergency. A defined intersectional approach to climate justice has emerged out of this context. Originally introduced by feminist legal scholar Kimberlé Crenshaw (1989) to 'conceptualize the way the law responded to issues where both race and gender discrimination were involved,' intersectionality has since been adopted by scholars and activists to analyze how environmental harm is intertwined with patterns of discrimination by race, class, gender, sexual orientation, and other social factors. This perspective is also reflected in organizational and interpersonal practices that seek to challenge hierarchies and systems of oppression both within the climate movement and across society more broadly (Federici, 2020).

Focusing on intra-societal injustice highlights the key role of intersectional factors—including race, age, class, disability, and indigeneity—in shaping how individuals and communities are affected by the climate crisis. Gender, for instance, is a critical determinant of climate vulnerability. The ecological and health burdens of flooding or water scarcity are often disproportionately borne by women due to gendered domestic responsibilities (Lama et al., 2021; Ajibade & Siders, 2021). Moreover, in many regions of the world, women are left to devise local adaptation strategies when men migrate to urban areas in search of work—thus assuming even greater responsibilities in agricultural work, water management, and household maintenance.

As Peter Newell (2005, pp. 72–90) argues, the global politics of environmental inequality 'reinforces and, at the same time reflects, other forms of hierarchy and exploitation along lines of class, race and gender,' resulting in what he terms 'double-discrimination.' A report by the UN Human Rights Council identified women, children, indigenous peoples, the elderly, and persons with disabilities in poor countries to be particularly vulnerable to such challenges (OHCHR, 2014). Likewise, in a 2019 UNHRC report on climate change and poverty, UN Special Rapporteur Philip Alston warned of the 'increasing risk of climate apartheid,' where the wealthy are shielded from climate impacts while the poor are left to suffer the consequences. This phenomenon transcends the Global North-South divide: for instance, research into Hurricane Katrina and the resulting displacement in the United States has shown that racial and socioeconomic inequalities—not merely natural factors—played a decisive role in determining who was most affected. Accordingly, the right to a safe, clean, healthy, and sustainable environment remains inaccessible to many around the world, particularly communities of color. Since 2009, the National Association for the Advancement of Colored People (NAACP)—the longest-standing organization linked to the US civil rights movement—has incorporated climate justice among its areas of advocacy. Through its Environmental and Climate Justice Program (ECJP), the NAACP mobilizes political and legal resources at local, state, and national levels to support poor and minority communities disproportionately affected by climate change (Sullivan, 2009).

Indigenous peoples are among the most severely impacted by intrasocietal injustice. While it is important to avoid homogenizing the diverse experiences of Indigenous communities, many are already bearing the brunt of climate change adversity. In the Amazon basin, for instance, climate-induced changes in flooding since 1999, have disrupted fish reproduction cycles, thus undermining the livelihoods of indigenous communities along the Amazon River (Kronik & Verner, 2010). In Asia, Typhoon Soudelor (2015) nearly caused the disappearance of the Wulai Tribe in Taiwan (Carling et al., 2015). In the Arctic, the Inuit are confronting rising food insecurity, thawing permafrost, and declining snow cover among other dire impacts (Atapattu, 2018).

Beyond the historical dispossession of their lands, Indigenous communities worldwide are persistently excluded from decision-making processes related to climate governance—both internationally and within their own countries. They are rarely involved in the development or implementation of mitigation and adaptation strategies that affect their territories (Müller & Walk, 2013). Within the United Nations Framework Convention on Climate Change (UNFCCC), some indigenous representatives may attend the negotiations, but their participation is largely symbolic, and their contributions are seldom reflected in the outcome documents. To counter this systematic exclusion, Indigenous peoples have organized international gatherings to develop shared positions and proposals within the UNFCCC processes. One such example is the Indigenous Peoples' Global Summit on Climate Change, held in April 2009, where participants adopted The Anchorage Declaration. This declaration acknowledged the climate crisis facing Mother Earth and affirmed the 'vital role of indigenous peoples in defending and healing Mother Earth' (Indigenous Peoples' Global Summit on Climate Change, 2009).

International Agreements and the Phantom of In/Justice

A review of the theoretical debates informing major global institutional meetings—including the 1992 Rio Conference, the 1997 Kyoto Protocol, the 2015 Paris Agreement, and subsequent UN Climate Change Conferences (COPs)—reveals that the legal and political complexities of climate justice are primarily situated at the international scale. These debates largely center on the historically constructed relationship between the so-called developing and developed states (Shue, 2016). A basic premise of climate justice discourse is that industrialized nations bear primary responsibility for climate change, whereas the populations most severely impacted—often in the Global South—have contributed least to its causes and possess far fewer resources to adapt to its effects. This stark asymmetry between the main perpetrators of pollution and its main victims raises profound questions of injustice that demand a reframing of justice so that it may encompass both spatial and temporal dimensions.

This international-scale dimension of injustice is historically rooted: it originates in colonial relations, becomes reinforced by globalization, and remains embedded in contemporary institutions. Normative claims in international injustice debates include demands that greenhouse gas emissions be drastically reduced; that adaptation and mitigation costs be equitably shared based on historic emission responsibilities (i.e. the polluter-pays-principle); that consumption patterns, especially in high-emitting countries, be transformed; and that more equitably global institutions be established. Moreover, the strong correlation between climate impacts and poverty suggests that justice in climate governance must go beyond the distribution of emission rights to also address the global distribution of wealth—natural, human, and social capital alike (Edenhofer et al., 2012).

Some of these normative claims found expression in the principle of 'common but differentiated responsibilities,' first articulated in the 1992 Rio Conference and five years later in the Kyoto Protocol. This period is especially significant for the evolution of the climate justice discourse. On the one hand, it saw the institutionalization of the Conference of the Parties (COP) system. As Imperatore and Leonardi (2023, p. 19) argue, the COP system from its inception 'appears as a strange hybrid between *political progressivism* (an unprecedented challenge is taken up, based on the recognition that the origin of atmospheric disruption is anthropogenic) and *economic neo-liberalism* (it is argued that only markets can solve the crisis: not public intervention, not international cooperation, not the participation of planetary civil society—-only markets).'

On the other hand, this period witnessed a rising mobilization of International NGOs such as the World Rainforest Movement, Friends of the Earth International, and the Third World Network. These organizations helped draw global attention to the local struggles of Indigenous and land-based peoples in the so-called Global South whose livelihoods were imperiled by escalating resource extraction enacted by neoliberal economic policies. Alongside other climate justice activists, they criticized the market-based emissions trading provisions introduced in the Kyoto Protocol. This collaboration led to the development of the Bali Principles of Climate Justice—a comprehensive 27-point program released in 2002 that aimed 'to build an international movement of all peoples for Climate Justice' (CorpWatch US et al.). This mobilization culminated in the formation of Climate *Justice Now! (CJN) network* during the 2007 Bali COP, uniting over 750 organizations across the Global South and establishing a communication hub to advocate for systemic transformation.

CJN called for a shift in responsibility onto those who have benefited most from 250 years of extractive development, rejecting the absurdity of the idea that market mechanisms could adequately address a crisis rooted in the very structures of global capitalism. The CJN network insisted that solutions to climate change must protect the most vulnerable, compensate the displaced, guarantee individual and collective rights, and respect the right of people to participate in decisions that impact their lives (Climate Justice Now, 2010). The Copenhagen Summit (COP15), however, marked a sobering turning point. Despite an overwhelming consensus on the urgency of climate action, negotiations eventually broke down, leading to widespread disillusionment among scientists, activists, and other observers. The failure of COP15 signaled not just a political impasse but a crisis of legitimacy for the international governance system as a whole.

The 2015 Paris Agreement was heralded as a breakthrough—a legally binding international treaty adopted by 196 parties at the UN Climate Change Conference (COP21). Its overarching goal was to hold 'the increase in the global average temperature to well below 2°C above pre-industrial levels,' while pursuing efforts 'to limit the temperature increase to 1.5°C above pre-industrial levels.' It also included a *Loss and Damage* provision acknowledging the need to address irreversible climate harms. For the first time in over 20 years of UN negotiations, a binding agreement brought all nations into a shared framework for climate action.

Yet the Paris Agreement represents a paradox: progress in formal inclusivity and ambition, but retreat in other areas that, on balance, indicate a long path ahead in terms of pursuing distributive and remedial justice. Enshrining the principle of voluntary action—a term that appears five times in the core text—the Paris Agreement shifts away from ideals of perfect justice (where actions are mandated and enforced based on principles of right and responsibility) toward imperfect ideals of what is good and appropriate, leaving implementation to the discretion of individual states.

A decade later, it is clear that climate policies, however reasonable in theory, face enormous challenges in practice. While world leaders in politics and business embraced each other and celebrated the signing of the Agreement with fanfare, the document itself—full of rhetorical flourishes and deliberate vagueness—fails to question the technocratic paradigm of unlimited growth underpinning the crisis. Notably, the word 'climate justice' is mentioned only once, and in inverted commas, as if signaling a discomfort or unease with its political implications. As Ghosh (2016) argues, this omission betrays a fear that acknowledging climate justice would challenge the existing balance of power, opening the door to demands for loss compensations (Ghosh, 2016).

From the signing of the Kyoto Protocol in 1997 to the Paris Agreement in 2015, COPs have repeatedly fallen short of the vision, impartiality, and ambition required to safeguard human and non-human development and well-being in the face of a deepening planetary crisis. Despite the growing urgency, the most recent COP meetings have often been characterized by incremental progress, half-hearted commitments, and a general inability to rise to the scale of the emergency. COP26 in Glasgow (2021) produced the 'Glasgow Climate Pact,' which pledged coal reductions and increased climate finance—but it was widely criticized for not going far enough in addressing the immediate need for deeper emissions cuts. At COP27 in Sharm El Sheikh (2022), the formal recognition of 'Loss and Damage' was hailed as a victory, yet the financial mechanisms set up to support vulnerable nations remain grossly inadequate. And at COP29 in Baku (2024), the climate finance deal was criticized by developing countries as woefully insufficient, due to its modest financial commitments, exclusionary negotiation process, and potential to exacerbate debt burdens.

These persistent shortcomings expose the structural inequities embedded in global climate governance and highlight the urgent need for more ambitious, inclusive, and redistributive climate finance solutions. As scholars have long insisted (Gardiner, 2011; Jamieson, 2010; Shue, 2016), addressing climate injustice requires multiple principles of justice—procedural, distributive, corrective, and intergenerational—to guide the design of institutional arrangements that can equitably manage the distribution of responsibilities, rights, and responsibilities across time and space.

Intergenerational Justice: Responsibilities Toward Future Generations

To date, most discussions on climate justice have been aligned with broader frameworks of global justice, albeit with differing methodological approaches. These perspectives converge around the principle that the burdens of climate change should be understood as global responsibilities. However, some theorists argue that global climate justice must not be confined to present-day inequalities; it must also encompass responsibilities toward future generations.

The adverse impacts of climate change, largely driven by human activity, pose serious challenges to justice and equity across temporal dimensions—between present and future generations. Decisions made today regarding climate change mitigation, adaptation, and resilience-building will profoundly shape the ability of people to effectively enjoy a clean, healthy, and sustainable environment now and in the future. In the case of many environmental challenges, the consequences of our present actions will only become fully apparent over decades—or even centuries. Some of the highest-risk impacts of climate change may not affect our children or grandchildren, but instead people born five, ten, or even twenty generations into the future. In a world of finite resources, the need to allocate these resources among competing interests cannot be ignored but must be faced transparently and ethically.

This proposition underscores the concept of intergenerational injustice—the growing ethical gap between those who cause climate harm and those who will bear its long-term consequences. According to Hiskes (2008) and Gardiner (2011), past and present lifestyles based on fossil fuel consumption and high greenhouse gas emissions have generated structural injustices that deprive future generations of the environmental conditions necessary for a flourishing life. To address this imbalance, some scholars advocate for constitutional environmental human rights that obligate current policymakers to take the interests of future generations into account, thereby enabling forms of intergenerational democratic representation, however indirect and incomplete.

While some theorists question whether duties can meaningfully be owed to individuals who do not yet exist—-arguing that future people cannot hold rights in the present—this ethical vision nevertheless underpins much contemporary climate activism. It forms the moral foundation of the protest movement initiated by Greta Thunberg, which gave rise to *Fridays For Future,* and subsequently to more radical movements like *Last Generation* and *Extinction Rebellion.* Thunberg, a Swedish teenager, rapidly emerged as a prominent voice in contemporary climate activism despite her youth and non-elite status. Her viral *Fridays for Future* campaign and sustained calls for intergenerational justice positioned her as a symbolic leader of a new youth-driven climate movement, particularly after 2019, when she galvanized global attention, and reframed the climate debate around the issue of equity. This term—equity—features prominently in her speeches, from her address to the European Parliament at COP24 in Katowice (2018) to the speech at the 2019 UN climate action summit that made her a household name..

While Thunberg is not the first youth climate activist, her radical method of protest created the 'Greta Thunberg effect': a global surge in youth activism and engagement among groups previously indifferent to environmental politics (Jung et al., 2020). As Han and Ahn (2020) observe, this movement represents one of the largest environmental mobilizations in history, with young people acting as agents of change and adopting the language of justice to critique the failures and inaction of political leaders and fossil fuel companies.

Thunberg and the wider movement she inspired have also exposed the limits of technocratic approaches to climate change favored by governmental and political elites. These institutional frameworks often rely on delay tactics, technical jargon, and promises of future technological solutions, effectively sterilizing the climate emergency through abstract managerial language (Smith, 2011). In more radical terms, *Last Generation* and *Extinction Rebellion* have appropriated the visual and symbolic power of museums and other art spaces to disrupt the conventional landscape of climate protests. By staging high-profile acts of dissent—such as gluing themselves to famous artworks or throwing paint at iconic paintings—they aim to provoke reflection on societal priorities: the preservation of cultural artifacts versus the preservation of the planet itself. Their strategic use of art within prestigious institutions underscores this stark tension and calls on the public to reconsider what is truly worth protecting (Villani, 2025).

Critics have repeatedly noted that the agreements and guidelines produced by successive COPs have largely preserved the economic logic of capital accumulation, sidelining justice-based concerns in favor of techno-scientific jargon. In stark contrast to this managerial approach, the new generation of climate justice activists has developed a narrative grounded in historical and structural injustice. As the popular metaphor suggests, while humanity may be facing the same storm, we are navigating in very different vessels—some in yachts, others in dinghies. Those in the former group are often more responsible for the crisis and better equipped to weather its consequences; those in the latter are more vulnerable and less culpable (Comparelli, 2022). This discomforting yet clarifying image has become common within activist discourse even as it remains conspicuously absent from the rhetoric of major governmental summits, where fundamental questions of systemic transformation are often elided.

It is thus not surprising that COP 28 (2023) was held in Dubai, a global hub of oil wealth, and that COP27 took place in Sharm El-Sheikh, both sites where civil society and activist pressure were silenced and marginalized. The symbolic choice of venues reflects the growing disjunction between elite climate change governance and grassroots movements, the latter increasingly driven by youth and by a vision of justice that challenges the status quo at its roots.

The Interconnectedness of All Beings: Cosmovisional Justice

To grasp the full scope of climate injustice, we must look beyond the disparities between countries and human populations and include the often-overlooked suffering of oceans, plants, and animals. Aside from the advocacy of a small group of vegan and animal rights activists, the plight of non-human life remains marginal in climate change discourse. Yet, it is crucial to acknowledge that while marginalized human populations typically have limited resources to adapt to climate change and exert little influence over decisions that affect them, non-human entities—including oceans, forests, and animals in industrial-scale farms, such as cows and pigs—have even less agency and virtually no representation in decision-making processes.

Addressing the full dimensions of climate justice requires recognizing that human beings are not isolated actors but deeply embedded in ecological systems. These ecosystems should not be viewed as passive backdrops or mere resources to be exploited and managed through a utilitarian logic (Haraway, 2016).

Over the past decade, a growing body of literature has reflected on the role of the non-human in climate ethics, exposing the anthropocentric assumptions underpinning mainstream notions of justice (Huntjens & Zhang, 2016). This critique has emerged across disciplines—from philosophy and anthropology to sociology and literary criticism—highlighting the deep entanglements between cultural worldviews, institutional power, and environmental degradation (Kohn, 2013; Viveiros de Castro, 2015). As Jacques (2006, p.85) puts it, dominant frameworks of ethics and governance still rest on a worldview that 'believes that humanity is totally independent of non-human nature and that moral obligation depends on relative and immediate human benefit, otherwise, ethics sees no obligation of human concern' (Jacques 2006, p. 85).

Among the most influential voices challenging this paradigm is Bruno Latour (1999, 2017, 2018), who argues that the Anthropocene requires a new political imagination that recognizes non-human beings as legitimate political actors. He proposes a 'parliament of things' in which humans are tasked with reconstructing political and epistemological relations to the Earth. Latour's call to transition from the 'Ancient' to the 'New Climate

Regime' reflects a shift away from the dualistic thinking that separates Nature and Society. The 'New Climate Regime' recognizes the deep interconnectedness between human and non-human forces, acknowledging that climate change is not an external threat but a condition deeply shaped by human systems.

Latour's new ontology calls for a more inclusive and responsible approach to global governance and action. He critiques the extension of nation-state hegemony over the Earth through an aggressive form of eco-modernization, inviting instead a return to the majesty of Gaia,[1] where the distribution of agency becomes a central political concern. This relational ontology recognizes the interdependence of all beings and encourages a form of multispecies justice that shifts the subject of justice from the individual to a vast ecological range of relationships that sustain life. It considers the capacity humans have to affect these relationships and their ability to understand and conceive of rights and responsibilities, and calls for a multispecies justice—a shift in focus from individuals to the dense ecological relationships that make life possible. As Celermajer et al. (2021) argue, such justice requires reimagining institutions and ethics to reflect the mutual responsibilities and vulnerabilities of all species.

In particular, more-than-human justice rejects three interrelated premises: (1) that human beings are physically separate or separable from the rest of nature; (2) that humans are uniquely endowed with mind, consciousness, and *agency*; and (3) that human beings are inherently more important than other beings (Srinivasan & Kasturirangan, 2016).

Indigenous communities have long criticized the anthropocentric frameworks embedded in sustainable development discourse and have consistently sought to resignify climate justice through their epistemologies and cosmologies, often imbuing the Eurocentric vision of climate justice with new connotations in order to raise their voices on climate change action. Although they are often identified as key stakeholders, Indigenous peoples are systematically excluded from decision-making and

[1] In his book Facing Gaia Latour (2017) states that to deal with the ecological challenge, we need a new paradigm of thought. This new 'worldview' is based on a perspective shift he partly already sees effectuated in the work of James Lovelock ([1979] 2009). A discussion of the implications of this new perspective brings Latour to a new concept of matter, an 'earthly materialism' that is compatible, he says, with an 'earthly humanism' that includes nonhuman entities in a large pluralist society and with a new reading of religion that also makes room for a pluralism of beliefs.

their knowledge systems are devalued compared to the 'scientific' knowledge produced in the Global North. This climate coloniality (Sultana, 2024) of 'expert knowledge' and epistemic inequality not only violates Indigenous rights but also deprives them of the plural perspectives and rationalities necessary to solve the climate crisis. To challenge this exclusion, Indigenous communities have launched various initiatives to articulate a shared cosmovision grounded in the defense and care of Mother Earth as opposed to a human-centered nation-state paradigm.

One landmark initiative was the *World People's Conference on Climate Change and the Rights of Mother Earth* convened by Bolivian president Evo Morales in Cochabamba in April 2010. Bringing together over 30,000 participants from civil society, social movements, and public institutions, the conference produced a comprehensive platform rooted in Indigenous principles of environmental harmony and anti-colonialism. Among its outcomes was the Proposal for a Universal Declaration of the Rights of the Mother Earth (Bolivia, Plurinational State of, 2010). This declaration affirms that Mother Earth is a living being with rights that must be legally guaranteed—a transformative vision that expands climate justice from an anthropocentric to a cosmocentric one. The declaration critiques carbon markets and the commodification of nature, particularly the UNFCCC's REDD (Reducing Emissions from Deforestation and Degradation of Forests) program, and calls for an International Climate and Environmental Justice Tribunal to judge and prosecute activities that harm the Earth (Anonymous, 2010).

Drawing on Boaventura de Sousa Santos' (2010) 'epistemology of the South,' this decolonial approach invites us to challenge the hegemony of the Eurocentric and çivilizing worldview, which has long posited itself as the sole horizon of progress and rationality. The cosmocentric vision of many Indigenous traditions—and now of an expanding group of scholars and activists—offers alternative ethical, legal, and political frameworks rooted in relationality, reciprocity, and respect for all forms of life. This perspective also reveals how the extractive capitalist economy, based on colonial patterns of land appropriation and resource exploitation, continues to drive the climate crisis and its unequal impacts. Extractivism disproportionately benefits nations, transnational corporations, and consumers in the Global North, while imposing ecological devastation and social dislocation on historically colonized communities in the Global South. The decolonial approach to the epistemology of the South, nurtured and adopted by several networks of Indigenous people, criticizes the anthropocentric rationalism that considers human beings the center of everything, and Mother Earth as just a commodity to be traded on the free market.

Ultimately, the cosmovision dimension of climate change recognizes the interconnectedness of all beings and the broader ecological and spiritual worldviews that shape the way injustices are perceived and experienced. This perspective transcends human-centered narratives, recognizing that the exploitation of the Earth, animals, plants, and even the cosmos itself is a form of injustice that affects all living entities. It challenges anthropocentrism by understanding that humans are not separate from but intricately connected to the natural world and its other-than-human inhabitants. From this standpoint, climate justice is not merely a legal or political obligation, but a moral imperative to tend to the systemic violence perpetrated against the land, non-human life, and the spiritual well-being of the planet. Such a cosmovision demands an ethical response grounded in reverence, interdependence, care, and shared responsibility for the Earth.

What Do We Talk About When We Talk About Mobility Justice

The concept of mobility justice has emerged as a vital framework for understanding the deeply uneven geographies of movement that characterize our globalized world (Sheller, 2018). At its core, mobility justice asks not only who is able to move, but also who is immobilized, under what conditions, and with what consequences. It interrogates the power relations, historical processes, and regulatory mechanisms that delineates who enjoys freedom of movement and who is subjected to barriers, surveillance, and exclusion. In doing so, mobility justice brings into sharp relief the ways in which the ability to move—or to remain in place—has become a central axis of social and environmental justice in the twenty-first century. As Mimi Sheller (2018) argues in *Mobility Justice: The Politics of Movement in an Age of Extremes*, the right to move—and the power to immobilize—are key justice concerns in our contemporary world. Human mobility is not a neutral or universal entitlement; it is structured through historical, political, and economic forces that distribute the ability to move—-or stay put—-along lines of race, class, gender, and nationality.

Mobility justice is not merely about the right to cross borders, but about challenging the systems that produce and legitimize these hierarchies of movement. It requires us to address the 'unequal relations of movement, dwelling, and belonging' that are rooted in histories of empire, race, and capital. It demands that we see the regulation of mobility—through passports, visas, carrier sanctions, and biometric surveillance—not as technical or administrative matters, but as central mechanisms of global injustice. In the context of climate change, pandemics, and intensifying border securitization, the struggle for mobility justice is inseparable from broader struggles for climate justice, racial justice, and the right to breathe and move freely on a shared planet.

In the United States,[2] the current migration regime provides a vivid case study for how mobility justice is both produced and denied through law, policy, and public discourse. In recent years, US migration policy has entered what many scholars and advocates describe as a constitutional crisis, marked by the criminalization of migrants and the militarization of borders. During Donald Trump's first term as president, migrants—particularly those from Latin America and the Global South—were routinely depicted as criminals, rapists, and terrorists. This rhetoric, amplified by executive orders and legislative actions, normalized the use of arrest, imprisonment, and deportation as tools of migration governance. The promise to build a border wall and the invocation of the 'Alien Enemies Act' of 1798 (which allows for the detention and deportation of so-called 'enemy aliens' without due process) further entrenches the legal and symbolic construction of migrants as invading enemies. By 2025, executive orders such as 'Protecting the American People Against Invasion' have extended these logics, with tens of thousands of migrants—including permanent residents—detained in facilities ranging from Guantanamo Bay to military bases on US soil, often without access to judicial review.

This escalation of enforcement and detention, including the use of military flights for deportations, exemplifies what Giorgio Agamben (1998) terms a 'state of exception'—spaces where normal legal protections are suspended and individuals are reduced to 'bare life.' The carceral

[2] The following section from Trump's first term until reflection of the power of the passport, draws on: Mimi Sheller, "Mobility Justice Now: Movement in an Age of Extremes," Keynote Talk for Re-imagining Climate (Im) mobilities in the Panicocene, Durham University, U.K., 8-9 May 2025.

turn in migration policy is not merely a response to contemporary crises: it is rooted in a long history of racialized and colonial control over movement. As Sheller (2018) and others have shown, the United States has a deep legacy of using legal and extralegal means to police the mobility of marginalized populations, from the Fugitive Slave Act to the internment of Japanese Americans, and now to the mass incarceration of Black, Latino, and undocumented communities (Mountz, 2011; Loyd & Mountz, 2014).

The US case is emblematic of broader global patterns, where the regulation of movement is fundamentally shaped by colonial histories. Colonial powers established systems of forced migration (slavery, indenture) and immobilization (pass laws, internal borders), which have been reconfigured in the infrastructures of the modern border regime. The invention of the passport, as John Torpey (2018) demonstrates, was intimately tied to imperial projects of population control and surveillance. Today, the passport index starkly illustrates these colonial continuities: citizens of former colonial powers and wealthy states of the Global North enjoy high mobility, while those from postcolonial and poorer nations face severe restrictions. The passport thus becomes a contemporary marker of global privilege and precarity, determining access not only to travel, but to safety, opportunity, and rights. The urgency of mobility justice today is thus inseparable from the legacies of colonialism. Colonial powers established global systems of racialized control over movement, forcibly displacing millions through slavery, indenture, and forced labor, while simultaneously restricting the mobility of colonized populations through pass laws, internal borders, and racialized policing (Shilliam, 2015; Sheller, 2020a). These colonial logics have not disappeared; rather, they have been reconfigured in the infrastructures of contemporary border regimes, visa systems, and citizenship laws. The very invention of the modern passport in the early twentieth century was bound up with imperial anxieties about controlling the mobility of colonial subjects and managing the flows of labor and capital (Torpey, 2018). Today, the passport remains a powerful technology of inclusion and exclusion, sorting the world's population into those with 'strong' passports, who can travel freely across borders, and those with 'weak' passports, whose mobility is severely restricted.

As we will analyze below, the passport index, which ranks national passports according to the number of countries their holders can enter without a visa, starkly illustrates these global inequalities.[3] It highlights how the different mobility regimes are not a neutral reflection of international relations, but a reproduction of colonial and neocolonial power structures. They draw a contemporary map of global privilege and precarity that shapes opportunities for travel and work, but also access to safety, family reunification, and the right to seek asylum (Figs. 2.1 and 2.2).

As climate change accelerates, the specter of 'climate migrants' is increasingly invoked to justify further militarization of borders and the expansion of detention infrastructures. These policies often frame climate-induced mobility as a threat to national security, rather than as a symptom of historical and ongoing injustices—including climate coloniality (Sultana 2022b). Yet, as Sheller reminds us, the problem is not migration itself, but the global capitalist and colonial system that underwrites vast social inequalities and structures of resource extraction. Climate mobilities are inherently relational, entangled with the movements of weather events, water, non-human species, and energy flows. Mobility justice therefore demands attention to the multiplicity of climate-related movement—encompassing not only migration, but also forced immobility, relocation, and the right to remain in place.

The struggle for mobility justice is fundamentally about the right to have rights—a principle famously articulated by Hannah Arendt and echoed in contemporary debates about citizenship, belonging, and planetary justice. In the United States and globally, the denial of mobility is a denial of personhood, rendering certain populations 'deportable aliens,' 'illegal aliens,' or 'criminal enemy aliens' (De Genova, 2002; Jones, 2017). The violence of borders—manifest in deaths at sea, in deserts, and in detention camps—is a direct consequence of policies designed to regulate and restrict movement, often in the name of security or sovereignty.

By foregrounding mobility justice, we can better understand how contemporary crises—whether climate-induced displacement, pandemic lockdowns, or border militarization—are not simply challenges of management or security. They are deeply political questions about who counts as a person, who is entitled to rights, and whose lives are valued or made disposable. In this way, mobility justice offers both a critical analytic

[3] https://www.passportindex.org/

Fig. 2.1 Antarctica World Passport Office (2017) forms part of Lucy + Jorge Orta's research project 'Antarctica,' which explores topics relating to the environment, politics, autonomy, habitat, mobility and relationships among peoples. For the artists, Antarctica represents Utopia: a continent whose extreme climate imposes mutual aid and solidarity, to research and collaborate together. The ice landscape is a filter for the kaleidoscope that makes up our nations and multiethnic identities, concentrating the colours, races and creeds into the sum of light, and the purity of hope. Reconditioned military trailer, painted steel, glass, wood dug-out canoe, canteens, linens, various found objects, Antarctica flag, online passport inscription, Antarctica World Passports, iPads, passport stamps, ink pads. Photo: Bertrand Huet. Courtesy Lucy + Jorge Orta/ADAGP Paris, 2025

Fig. 2.2 Anneli Skaar, Nansen's Pastport, 2020. Published by Two Ponds Press. Nansen's Pastport is an artistic re-invention of Norwegian polar explorer and humanitarian Fridtjof Nansen's refugee passport for post-Great War Europe. The book, designed and conceived by artist Anneli Skaar is re-imagined as a climate refugee passport for humanity, using Nansen's own words from his 1922 Nobel Lecture in a contemporary context, inviting consideration on how to meet the pressing issues of current and future climate migration with wisdom from the past. Courtesy of the artist Anneli Skaar

and a call to action: to dismantle the colonial and racialized architectures of movement that divide humanity, and to imagine new forms of solidarity and planetary belonging.

The Mask and the Wall: The New Necropolitical Frontier

As we have seen, climate justice is understood in a multiplicity of ways; it functions as a mirror reflecting the many visions of justice across disciplines, regions, and bodies. As Verlie and Neimanis observe: 'we are the climate change we witness' (2023, p. 118). There is no singular climate change, they argue, just as there is no omniscient witness. Building on the insights of feminist, decolonial, and disability justice scholars, they conceptualize climate change as a pervasive and unequal planetary crisis that affects the very essence of life, which is reflected in our personal entanglements with it. 'We are all breathing the climate crisis together, although differently, according to our different forms of embodiment, our different socio-political situations, and our different historical-geographical locations' (2023, p. 123). From this perspective, the climate crisis must be recognized as deeply personal and unevenly distributed—a present-tense emergency that demands an inclusive ethics of witnessing attentive to the many dimensions of injustice it encompasses.

In their *Breathing Climate Crises*, Verlie and Neimanis invoke *The Universal Right to Breathe* published by Mbembe in 2020, a year that witnessed multiple interrelated respiratory injustices: the intensifying climate crisis, the Covid-19 pandemic, and racist police violence, encapsulated in the Black Lives Matter slogan 'I can't breathe' (Hardeman & Karbeah, 2020; Habtom & Scribe 2020). The year 2020 made painfully visible the convergence of capitalism, settler colonialism, mass extinction, heteropatriarchy, ableism, and climate catastrophe. As Mbembe (2020) writes: 'Before this virus, humanity was already threatened with suffocation. If war there must be, it cannot so much be against a specific virus as against everything that condemns the majority of humankind to a premature cessation of breathing, everything that fundamentally attacks the respiratory tract, everything that, in the long reign of capitalism, has constrained entire segments of the world population, entire races, to a difficult, panting breath and life of oppression. For the Cameroonian postcolonial intellectual, the 'right to breathe' transcends the biological act; it becomes a fundamental right to existence: 'It is an originary right to living on Earth, a right that belongs to the universal community of earthly inhabitants, human and other.' Stressing that humankind and biosphere are one—'alone, humanity has no future, [because] each of us belongs to the same species, that we have an indivisible bond with all life'—Mbembe sheds light on how modernity has waged an interminable war on life. For him, the COVID-19 pandemic represented a moment to rethink the damage we, as humans, have inflicted on the Earth's lungs and on the planetary body itself.

In the same year, the philosopher Paul B. Preciado's essay *To Learn from the Virus* (2020) reflected on the biopolitical and territorial reconfiguration emerging from the pandemic. Having personally recovered from COVID-19, Preciado explored the virus' reshaping of border regimes. As governments scrambled to contain the virus, they enacted border closures and lockdowns that halted international mobility to control the outbreak, protect public health, and prevent healthcare systems from being overwhelmed. Global tourism plummeted by over 70% in 2020 (WTO, 2023), while millions who relied on cross-border travel for work, education, or family reunification faced profound challenges. More generally, the situation highlighted the interconnectedness of global mobility and the vulnerabilities within existing immigration systems.

At the same time, we are less aware (or at least less concerned) about the fact that, for millions of people of the Global South, borders are deadly walls that chain would-be migrants or asylum seekers to their place of origin. Stressing the tight relationship between immunity and border politics, Preciado writes that the virus is always the alien, the other, the foreign intruder. Border politics, long predicated on fear of the Other, mirror the way the (corona-) virus framed strangers as potential 'others.' Human mobility and viral contagion are feared as conflatable entities. Fear, in such settings, materializes in the form of material borders being imagined, re-established, or re-activated (Fanon, 2001).

Indeed, Preciado argues, recent decades have seen a redefinition of state sovereignty along neocolonial and identitarian lines, driven by a return to the physical border as the locus of national integrity. 'Using techniques of exclusion and death,' he writes, 'European and North American societies have decided to construct themselves as entirely immunized communities, closed to the east and the south, even though these two regions are its chief suppliers of fossil fuels and consumer goods.' The construction of this political immunity exemplifies the neo-sovereignist governmentality across geographies: Europe closed its borders in Greece, Italy, and Spain in 2015 and built the largest outdoor detention centers in history around the Mediterranean. The externalization of borders is a process that started in Australia two decades ago and is now a common practice adopted by the United States and several European countries (Dastyari, Nethery, Hirsch, 2023). The destruction of Europe paradoxically began with the construction of an immune European community, open in its interior but completely closed to foreigners and migrants.

At the same time, Preciado argues, what is now being tested on a global scale through the management of COVID-19 is a new way of understanding sovereignty: 'The body, your individual body, as a life space and as a network of power, as a center of production and of energy consumption, has become the new territory where the violent border politics that we have been designing and testing for years on 'others' are now expressed, now taking the form of containment measures and of a war against the virus.' In Foucault's terms, the COVID-19 epidemic radicalizes and shifts biopolitical techniques to the scale of the individual body while generalizing mechanisms of immunization and control previously applied primarily to racialized and excluded populations at the borders of national territory.

Preciado's argument culminates in a powerful metaphor: 'The new necropolitical frontier has shifted from the coast of Greece toward the door of your home. Lesbos now starts at your doorstep. And the border is forever tightening around you, pushing you ever closer to your body. Calais blows up in your face. The new frontier is the mask. The air that you breathe has to be yours alone. The new frontier is your epidermis. The new Lampedusa is your skin.' For years migrants and refugees have been treated as infectious threats, confined to detention centers—political limbos where they remained without rights and citizenship in perpetual waiting rooms. During the pandemic, many experienced a similar suspension in the detention centers of our own homes.

By foregrounding the mechanisms through which some lives are rendered disposable, Mbembe critically interrogates the geopolitical and biopolitical forces shaping contemporary modes of existence and governance. His concept of the 'necropolitical frontier' reveals how mechanisms of control, such as passports and racial classifications, create spaces in which lives are regulated through systems of exclusion and disposability. As analyzed above, the passport, as a tool of state power, becomes a marker of inclusion or exclusion, determining who has the right to mobility and survival. When these bureaucratic tools intersect with racial hierarchies, they delineate populations deemed 'worthy' of protection while relegating others to spaces of abandonment or precarity. In this sense, the necropolitical frontier (Mbembe 2019) is both a physical and symbolic boundary, where the intersection of race and mobility regimes enforces structural violence and stratifies humanity into categories of those who live and those left to die: 'Human bodies are increasingly divided between those that matter and those that do not, those who can move and those who cannot or should not or should only move under very strict conditions' (Mbembe, 2019, p.11).

Understanding the necropolitical frontier means recognizing the liminal spaces where power—state or non-state—produces vulnerability and enforces violence and exclusion. In today's globalized world, freedom of movement is a key marker of social stratification. According to Zygmunt Bauman (2013), mobility has ceased to be a universal entitlement; rather it is a privilege that is distributed disparately across social and economic hierarchies. Those who possess this privilege enjoy the right to move freely—for work, family, study, or pleasure—while the marginalized are often immobilized by borders, surveillance, and restrictive migration policies confining them to precarious spaces. In this way, mobility becomes not only a symptom but a reproducer of global inequalities, upholding structures of neoliberalism and dynamics of disposability.

Who Counts as a Person? Mobility Regimes and Social Justice

In the following chapters, we will delve into how migration, traditionally perceived as a problem, is increasingly becoming a critical concern in discussions on climate justice, especially in the context of climate change and environmental degradation. The effects of climate change and natural hazard-induced disasters are exacerbating existing vulnerabilities and reshaping global migration patterns in complex and multifaceted ways. Both the sudden and gradual impacts of climate change profoundly influence human (im)mobility. As a result, the movement of people—both within and across borders—has been and will continue to be, substantially re-routed by these transformations. For some, migration functions as an adaptive response to climate change; for others, movement is not possible, and so they remain immobile. This contrast illustrates the wide array of mobility outcomes and reveals the challenges of analyzing the intricate dynamics of different mobility regimes. The interdisciplinary concept of mobility regimes express the intersection of social, economic, political, and technological forces that shape mobility in society (De

Genova, 2002; Squire, 2012). Shedding light on how mobility is not just a matter of physical movement but also a question of equity, access, and rights, the concept of mobility regimes raises important questions about justice. It helps expose how mobility is not merely a physical process of people crossing borders but is intricately linked to justice insofar as it is deeply tied to systems of privilege, inequality, and systemic injustices that can affect the very ability to move and the conditions under which migration occurs.

As Saul and McAdam (2008, p.3) ask, are climate-induced (im) mobilities to be viewed as a 'refugee issue, a migration issue, a human rights issue, an environmental issue, a security issue or a humanitarian issue?' (p. 3). To begin answering this question, we must first ask: who counts as a migrant?

The United Nations (UN) defines a migrant 'as a person who moves to a country other than his or her usual residence for a period of at least a year'.[4] However, in our times, this definition encompasses a broad range of circumstances: individuals who move to another city or state for work; those fleeing war, political oppression, or famine; students participating in an Erasmus program; people pursuing family reunification; and those displaced by environmental events such as floods or hurricanes.

Before exploring the issue of climate-induced (im)mobilities, it is crucial to consider two key components of international migration and travel: mobility regimes and the visa system.

Visa systems serve as the gatekeeping mechanism through which countries regulate entry based on nationality, purpose of travel, and duration of stay. Different visa categories—tourist, work, student, transit—carry varying conditions and restrictions. The requirements for obtaining a visa vary widely and reflect a country's diplomatic relations, economic priorities, and security concerns. Yet access to mobility is far from equal: while many citizens of countries in the Global North enjoy visa-free travel or streamlined entry to multiple countries, individuals from the Global South frequently face stringent requirements, including proof of financial stability and return guarantees. These discrepancies underscore the deeply uneven power dynamics that visa systems perpetuate.

[4] https://www.un.org/en/global-issues/migration

Visa restrictions, immigration policies, border enforcement practices, and refugee resettlement policies are heavily influenced by colonial legacies and global power dynamics, with wealthy countries often imposing harshly restrictive policies on the citizens of poorer nations. This creates an inequitable global mobility system where citizens from wealthier countries can move freely, while migrants from the Global South face barriers to entry. As Saskia Sassen (2012) has shown in her work on global cities and migration, mobility regimes tend to privilege specific groups (e.g., tourists, business elites, skilled workers) while restricting the movement of others, especially refugees and undocumented migrants.

Migrants from marginalized backgrounds—particularly racialized minorities, women, and low-income workers—often encounter systemic barriers to mobility such as discriminatory immigration policies, exploitation in labor markets, and restricted access to social services. For instance, undocumented migrants are often excluded from legal protections and are vulnerable to exploitation and abuse. According to Fanizza and Omizzolo (2018), low-wage labor migrants in the informal economies are subject to forms of economic exploitation that reproduce broader patterns of class- and race-based injustice. Migration, in this sense, is not merely about crossing borders; it is about negotiating power structures that determine who is permitted to migrate, who is allowed to stay, and who is denied rights altogether.

As with climate (in)justice, mobility justice plays out across different scales, revealing contradictions between the national and the international. While there are legal frameworks and international instruments inspired by cosmopolitan ideas of justice, states still control their borders and the movement of people across them. The United Nations Refugee Convention (1951) and its 1967 Protocol, for example, establish a legal framework for refugee protection, yet in practice, the treatment of refugees frequently deviates from international standards. This discrepancy arises as countries often prioritize national security and economic interests over human rights. More broadly, the injustices inherent in current mobility regimes become pronounced when examining the rights of displaced individuals—those escaping conflict, environmental disasters, or persecution. According to the United Nations High Commissioner for Refugees (UNHCR), there are currently over 100 million displaced people worldwide, many of whom face significant barriers to mobility due to restrictive immigration policies. These barriers to movement can be understood as a form of global injustice, where the right to mobility is denied based on nationality, ethnicity, or social status.

At the heart of these contradictions is the nation-state, the principal arbiter of who may move and who must be walled up by the necropolitical frontier. The rights of non-citizens are subordinate to those of citizens, enshrining a hierarchical distinction in the allocation of rights. In this context, justice translates to honoring migration as a fundamental human right and involves the responsibility of states to uphold these rights. As Derrida (2000) famously observed, this contradictory logic manifests in the tension between the hostility that comes from border politics and the desire to give unconditional hospitality—an in-between condition he terms 'hospitality.' Here, hospitality is caught between open-ended moral principles and bounded, territorial laws. In today's global order, individuals categorized as tourists, experts, entrepreneurs, or expatriates experience enhanced mobility, benefiting from privileged legal and social status. Meanwhile, those defined as asylum seekers or economic migrants—often poor and in search of refuge—are blocked by fences, detained in camps, or forcibly deported. Birthplace and class shape one of the most glaring inequalities of our time: 'the unequal distribution of the right to emigrate' (Wihtol de Wenden, 2015, p. 49).

The so-called migration crisis in Europe, along with the displacement of populations from Syria, Afghanistan, Palestine, and other conflict zones, are stark reminders of how global mobility regimes fail the most vulnerable. Governments in the Global North—such as the US, Australia, the UK, and EU member states, have been widely criticized for their inadequate and often inhumane responses: detaining refugees and keeping them in degrading conditions, denying asylum on arbitrary grounds, and outsourcing border enforcement in line with political or economic interests.

To grasp the deep entanglement between mobility regimes and social justice, consider the ease with which an Italian passport-holder like the authors of this book (white, wealthy, etc.) can purchase a plane ticket and travel to Australia or the United States on the one hand, and, on the other, the life-threatening journey faced by an irregular Black migrant from a poor country who may spend years in transit enduring violence, abuse, and invisibility before reaching their destination. To understand the contradictions inherent in what we can call the 'moral geography of the world' (Musarò & Parmiggiani, 2022), one need only look at the passport index, which compares the travel possibilities of the world's various passport holders (Giacomelli, 2023; Walker & Giacomelli, 2024). While an Italian passport (like that of many European citizens) offers access to 188 countries, a passport from Iraq, Afghanistan, Syria, Somalia, Yemen, and Pakistan only allows access to 33. The exclusion is not personal but political—based on origin, not identity. As Mauro (2018) notes, it is not a sin of behavior, but a sin of origin (Giacomelli & Walker, 2021). The index comparing the value of passports tells us of the many countries where it is impossible to obtain a visa to expatriate since this visa is linked to relations between states—and therefore has a geopolitical value not linked to the person. Furthermore, this perspective reveals how the selection of migrants is typically based on three key criteria, which Ambrosini (2020) refers to as the 'three P's': passports, portfolios, and professions.

With respect to passports, for example, in recent years an undeclared migration policy has been put into practice that has granted millions of Eastern Europeans the freedom to circulate and seek work in the most prosperous and labor-hungry countries, including Italy. Thus today, out of 38 million immigrants in the European Union (EU), 17 are citizens of a member country, mostly from the East of Europe. Moreover, through different visa policies, the entry of (formally tourist) citizens of an increasing number of non-EU countries (e.g. Argentina, Brazil, Albania) has become accepted.

The global passport rankings, with the United Arab Emirates at the top, show that wealth is increasingly associated with enhanced mobility and privileges, a phenomenon often referred to as 'wealth whitening.' This trend is similarly reflected in the growing number of EU governments that readily allow the settlement of foreign investors, sometimes even granting them citizenship directly, as seen in countries like Cyprus and Malta. This practice effectively establishes a sort of *ius pecuniae*, or 'right by wealth,' where citizenship can be acquired through financial means, regardless of the ongoing public debates around ius *soli* or *ius sanguinis*. Meanwhile, public discourse and political rhetoric continue to focus on asylum seekers, advocating for their exclusion from EU territory by outsourcing border control to third countries such as Turkey, Niger, Tunisia, Libya, Albania, and Rwanda—where conditions are often inhumane and legally ambiguous. The disparity between public perception and reality drives European public discourse and the populist narratives that support it, reducing migrants from persons into bodies, from bodies into mere numbers, from numbers into waste, and from waste into surplus.

As Arendt (1966) warned, this process strips people of the protections afforded by citizenship, leaving them rightless and treated as the 'scum of the earth.'

Beyond the Residentialist Approach

Since migration is a major driving force of planetary and population health, it is necessary to analyze human mobility and climate change within a shared analytical framework—one that regards mobility as the norm rather than the exception. To better capture the complexity of our mobile world, we must move beyond national scale perspectives and the assumption that sedentary lives are the standard. Moreover, we recommend incorporating the socially transformative potential of migration into policy and planning initiatives designed to enhance long-term climate resilience and adaptation.

A justice-oriented approach to migration demands a critical reassessment of current policies and the recognition of migration as a fundamental human right, one that should not be constrained by race, class, nationality, gender, or environmental conditions. Achieving this requires contesting dominant assumptions about the nature and significance of migration.

As many critical migration studies scholars have observed (Balibar, 2002; De Genova, 2017), the term *migrant* has become a toxic term saturated with stigma. Malone (2015), for example, demonstrates how the term has evolved from a neutral descriptor into a dehumanizing label. Following this view, Bridget Anderson (2017) notes that the term *migrant* is not reducible to a marker of persons crossing international borders: 'migration'—she argues—'signifies problematic mobility'—a categorization that is always already embedded in regimes of control. Accordingly, 'not all mobility is subject to scrutiny, but "migration" already signals the need for control and in public discourse is often raced and classed' (Anderson, 2017, p. 1532). The regulation of migration, particularly the mobility of economically disadvantaged populations, is mediated by laws governing citizenship and concepts of national belonging. This historically and geographically specific construction of certain mobilities as 'migration'offers a critical lens through which to examine changes in migration policies as well as transformations in border and citizenship regimes. In this spirit, Anderson (2017, p. 1535) calls for turning 'the problematization of migration into a tool for inquiry.'

In the journal article *Who is a Migrant? Abandoning the Nation-state Point of View in the Study of Migration*, Scheel and Tazzioli (2022) build upon Anderson's call of 'problematizing the problem of migration' by posing the deceptively simple yet profound question: who is a migrant in Europe today?

Their analysis critiques what Favell (2007, p.271) termed 'the nation-state point of view of spatial mobility,' which continues to frame migration as a problem requiring constant monitoring as well as governmental interventions of regulation and control. Accordingly, 'this nation-state point of view is carried by dominant understandings of migration as movement to and residence in nation-state B from nation-state A. The letter informs policymaking as well as statistical and academic knowledge production on migration' (Scheel & Tazzioli, 2022, p. 2).

The concept of methodological nationalism conceives of societies as contained within national borders (Wimmer & Schiller, 2003), resulting in a widespread understanding of migration as the transfer of people between states. Yet borders are social and political constructs (Zanfrini, 2016), upheld through both material and symbolic practices. Border control manifests not only through walls, fences, and checkpoints but also through the externalization of borders to third-party countries. At a more abstract level, borders are 'zones of cultural production, spaces of creation and violation of meanings' (Wilson & Donnan, 2000, p. 64). Through such processes of symbolic bordering, nations draw lines between 'us' and 'them,' defining belonging and identity. These symbolic boundaries are reinforced through cultural, political, and social narratives that determine who is considered part of the national community and who is left out. This way, border control becomes not just a physical practice but a deeply embedded mechanism for constructing and maintaining national identity and sovereignty (Mezzadra & Neilson, 2013).

As Khosravi (2011) succinctly puts it: borders make migrants. Without borders, there would be neither migration nor migrants, only mobility and people on the move. Similarly, De Genova notes that it is 'the bordered definition of state territoriality that constitutes particular forms and expressions of human mobility as 'migration' and classifies specific kinds of people who move as 'migrants.' (2013, p. 253).

Returning to Saul and McAdam's question of how to conceptualize climate-induced (im)mobility, it becomes clear that to articulate the relationship between climate and mobility justice, we must move beyond the nation-state-centered conceptions of both. This 'container thinking' fails to account for the global turbulence—political, economic, and social—that drives contemporary patterns of mobility. Over two decades ago, Nikos Papastergiadis (2000) described *The Turbulence of Migration*, urging the social sciences to adopt a more dynamic view of culture and the 'social'—one willing to incorporate cultural differences outside traditionally bounded units of classification.

Today, in a world marked by globalization and the compression of space and time, cosmopolitan perspectives are more relevant than ever. As Ulrich Beck and Natan Sznaider (2006) observed, thanks to the dynamics of mobility, in combination with the advances in communication technologies, we are witnessing the 'cosmopolization' of the world. The global other is now in our midst. Cosmopolitanism, in Beck and Sznaider's reflections, allows one to defamiliarize oneself—as well as others—connecting individuals and groups in novel ways. Cosmopolitanism was not merely a normative ideal, as imagined by Kant, but an insistence on the development of cosmopolitan rights, laws, and institutions as the ultimate end of human evolution. Though frameable as a worthy aspirational ideal and noble pedagogic tool, it was also already an actual feature of everyday life. Yet its significance remains obscured by nation-based modes of seeing and setting priorities. Data collection, policy frameworks, public discourse, and conceptual tools for analyzing social change, still operate within national logics, even as global flows transcend them.

As Papastergiadis (2012) notes: 'Cosmopolitanism usually refers to the social transformation that arises from the mixture of different cultures. In the context of the political impact of migration, these signs of mixture are often interpreted through a kinetophobic prism' (p. 135). Such approaches fail to recognize that modern life is largely shaped by movement, and struggle to capture the transnational flows shaping contemporary societies. Failure to investigate how human mobility promotes cultural exchange while also leading to challenges like identity crises, cultural hybridization, and societal anxieties over integration and belonging, risks reducing migrants to threats against national identity or cultural cohesion (Anderson, 2016). Thus, methodological nationalism tends to obscure the various practices of bordering and boundary-making that construct, problematize, and target certain individuals as migrants. In contrast, embracing what Beck (2006) terms 'methodological cosmopolitanism' allows us to conceptualize new forms of belonging (Appadurai,

1990; Calhoun, 2008; Chouliaraki, 2013) and new visions of global justice (Appiah, 2006).

If climate justice 'is a framework that brings into view the intersection between climate change and the way social inequalities are experienced as structural violence' (Porter et al., 2020, p. 293), it is essential to integrate considerations of mobility justice into that framework. As Mimi Sheller argues, mobility justice interrogates 'the ontological definition of *who counts as a person*, which as we know has historically excluded women, the enslaved, queers, and the differently abled, and continues to exclude non-human animals, plants, and living entities that are included in many Indigenous ontologies' (Sheller, 2018, p. 39). In her seminal book *Mobility Justice* (2018), the sociologist posits that climate change constitutes a mobility crisis in at least three distinct ways. First, transportation-related emissions are a major contributor to global warming. Secondly, the built environments and land use patterns that support current mobility systems degrade ecosystems and intensify climate vulnerability. Finally, climate change displaces populations and disrupts infrastructure systems and supply chains globally, leading to environmental migration in the face of ecological and societal collapse (Sheller, 2018, p. 4).

Together with the new mobilities paradigm (Sheller & Urry 2006), which emerged from the 'spatial turn' in the social sciences (Lefebvre 2000; Sassen, 2007; Urry, 2002), Sheller's perspective highlights the constitutive role of movement in shaping social institutions and practices. We propose shifting analytical attention away from static territorial units (nation-states, societies) or isolated actors (individuals, groups, communities) to the power structures that regulate mobility: from border architectures to transportation infrastructures and the multi-scalar governance systems that structure circulation and containment.

The mobilities paradigm challenges the assumption that mobility is exceptional and sedentarism is the norm (Adey 2006; Schapendonk et al., 2021). Central to this approach is the notion that the different forms of mobilities—of people, materials, ideas, technologies, knowledge, and risks—are already central to 'producing and reproducing social relations on local, regional, and global scales' in our globalized world (Wiegel et al., 2019, 4). The mobilities paradigm extends beyond mere analysis of the movements of people and objects, encompassing the complex interplay between these movements and the asymmetrical power structures that both shape and are reinforced by them. Through this lens, climate mobilities are not aberrations or anomalies but deeply embedded within the histories and geographies of mobility control. This perspective requires a nuanced understanding of climate-induced movement as part of a broader, evolving socio-political landscape of mobility and stasis.

Conclusions

In this chapter, we have foregrounded the pursuit of social justice by examining how the interconnected global phenomena of climate change and migration intensify existing legal, economic, and social inequalities. In doing so, we stressed their profound implications for human rights and the broader framework of justice. To this end, we explored different theories of justice to offer a diversity of perspectives on climate change and migration through national, international, intergenerational, and cosmovision-based lenses. While each approach has its limitations, they nonetheless provide valuable tools and policies to address climate change and migration.

Our analysis revealed that both climate and mobility justice are interpreted in varied and often contested ways (Neef et al., 2024). These competing frameworks reflect deep-seated tensions between national interests, economic disparities, and structural power dynamics.

As we saw when analyzing the array of existing approaches, climate justice 'recognizes humanity's responsibility for the impacts of greenhouse gas emissions on the poorest and most vulnerable people in society by critically addressing inequality and promoting transformative approaches to address the root causes of climate change' (Meikle et al., 2016, p. 497). Thus, despite the heterogeneity of perspectives on climate justice, these frameworks share at least one thing in common: they focus on the equity and justice aspects of both the causes and the effects of climate change. These 'root causes'—rarely named explicitly—lie in the capitalist system and the colonial legacies of its extractivist model, with the attendant understanding that poverty and power imbalances act as 'multipliers' for the negative impacts of climate change.

Through our examination of the historical and political evolution of global climate governance frameworks, we have argued for the adoption of an intersectional lens. Industrialized nations in the Global North, having historically relied on carbon-intensive industrialization to drive economic growth, now impose climate mitigation imperatives that constrain the development of poorer nations. Countries in the

Global South—particularly low-income and emerging economies—are now constrained from pursuing similar developmental trajectories under current climate imperatives (Moellendorf, 2014). Compounding this inequity, many Global South countries are disproportionately affected by the climate crisis, enduring extreme weather events, recurrent flooding, and prolonged droughts. This underscores a fundamental inequity wherein affluent nations bear disproportionate responsibility for historical emissions, while poorer nations face disproportionate harm while lacking the resources to adapt to the impacts of climate change.

We have also highlighted how intra-societal injustice and intersectionality further impact the degree of vulnerability to climate change, which falls overwhelmingly on low and lower-middle-income countries. Vulnerability to risk, as well as intensities of suffering, are determined by levels of economic development rather than the climate per se. Engagement with intersectionality draws out the interlocking systems of oppression and exploitation that shape vulnerabilities. As Sultana points out: 'Vulnerable communities are not homogenous; instead, they are gendered intersectional (by class, race, Indigeneity, migrant status, etc.)' (2022, p. 3). Thus, it is essential to acknowledge the intrastate variations in the impact of the climate crisis. Notably, Indigenous communities are disproportionately affected by the adverse consequences of climate change, highlighting the intersectionality of environmental and social justice issues within the broader context of climate resilience and adaptation strategies.

The inadequacy of the 2015 Paris Agreement—its failure to implement legally binding and enforceable mitigation targets—has prompted climate justice movements to expand their focus. Activists are now demanding not only equity in emissions reductions, but also in participatory governance based on comprehensive information, access to judicial remedies, and compensation for loss and damage. The Paris Agreement offered a rhetorical nod to many of the chief climate justice concerns but fell far short of satisfying the substantive concerns of climate justice advocates.[5] The Agreement's Preamble acknowledges values such as the 'principle of equity,' the reality that some people are 'particularly vulnerable' to climate impacts, the rights of Indigenous peoples, migrants, and children, as well as principles of 'gender equality,' 'intergenerational equity,' and

[5] Arguing that the agreement imposes unfair burdens on the American economy, president Trump is withdrawing from the Paris Agreement (again), thus reversing U.S. climate policy.

the 'imperatives of a just transition of the workforce.' But this agreement, like the ones following it, remains silent about the fact that the institutions and economic policies responsible for climate destabilization are also the underlying causes of economic inequality.

In this context, we critiqued the prevailing market-driven solutions. As observed in the climate policy negotiations on *Loss and Damage*, the current development paradigm treats economic growth as limitless and as an end in itself. It is widely believed to bring cascading benefits such as poverty alleviation despite evidence showing that global economic growth has, in fact, only exacerbated inequalities (Alvaredo et al., 2018). This paradigm prioritizes economic growth over social and cultural inclusion, environmental protection, biodiversity conservation, and ecosystem restoration, and silences holistic approaches that consider the interconnectedness of life systems. Consequently, it is contradictory to assume that the market can resolve the problems it has created. Why would it, especially since doing so would come at its own expense?

We also explored how migration is increasingly central to climate justice debates. Climate change-induced displacement—both sudden and slow-onset—requires urgent policy attention. Yet migration is still widely treated as a problem to be solved rather than as a site of justice from which to rethink existing policies beyond the nation-state-centric approach. We must then challenge the mainstream understanding of migration and denounce the unequal distribution of the right to emigrate. Although the right to mobility has been theorized since the seventeenth century by philosophers and jurists such as Grotius, Locke, Voltaire, and Kant and claimed by numerous transnational movements as a prerequisite for world citizenship, the right to enter another country remains the privilege of those at the top of the social hierarchy. Today, this right is effectively reserved for those with social and economic capital: the tourist, the expatriate, and the international student. In contrast, those forced to migrate due to conflict, danger, poverty, anguish, or climate collapse are often constructed as burdens or threats—a social dead weight to be rejected as a 'social problem.'

Border control systems exacerbate vulnerabilities and reinforce social stratification, with the affluent enjoying mobility and the marginalized confined by restrictive policies. As Mbembe (2019, p. 11) states when recalling the systemic violence of the necropolitical frontier: 'Only what can potentially generate value counts as life. In this context, borders are meant to concretize the principle of dissimilarity rather than that of affinity. They are not only obstacles to free movement. They are boundaries between species and varieties of the human.' Nevertheless, far from recognizing migration as a fundamental human right, free from race, class, nationality, gender, or environmental status, the term 'migrant' has shifted from a neutral term to a stigmatizing and dehumanizing categorization of others.'

Finally, drawing attention to how climate justice and mobility justice have historically been siloed due to their distinct focus areas, this chapter argued that addressing these interconnected challenges requires a unified approach. It is essential to uncover deeper inequalities and unveil alternative views that could benefit both people and the planet.

Rather than starting from the sedentary perspectives of nation-states, societies, or individuals, we emphasized the need to focus on the power structures that govern mobility, immobility, channels, and borders across various scales. Applying the new mobilities paradigm (Sheller & Urry, 2006) to climate mobilities, we critically explored how the right to dwell and move is regrettably demarcated by climate mobility regimes, especially for populations that have contributed the least to climate change. The concept of mobility regimes highlights that movement is not just about physical relocation but also about equity, access, and rights. It is closely tied to justice, as people's ability to move is shaped by systems of privilege, inequality, and systemic inequities that influence both the opportunity to migrate and the conditions under which migration occurs. Such an alternative framing of mobilities serves to disrupt the dominant research paradigm, which is underpinned by entrenched 'sedentary epistemologies' (Sheller, 2020b, p. 20). Whyte (2016) critique sedentary epistemologies as integral components of colonial knowledge systems, legal frameworks, land tenure regimes, economic structures, and socio-political institutions. These epistemologies have historically contributed, and continue to contribute, to processes of marginalization in the past and present. This perspective aligns with broader critiques of epistemic coloniality, which highlights how Western knowledge systems have been privileged and imposed upon the world at the expense of indigenous and local knowledge, leaving enduring hierarchies and perpetuating intellectual dependencies.

References

Adey, P. (2006). If mobility is everything then it is nothing: Towards a relational politics of (Im)mobilities. *Mobilities, 1*(1), 75–94. https://doi.org/10.1080/17450100500489080

Agamben, G. (1998). *Sovereign power and bare life.* Stanford University Press.

Ajibade, I. J., & Siders, A. R. (2021). Introduction: Climate change and planned retreat. In I. I. J. Ajibade & A. R. Siders (Eds.), *Global views on climate relocation and social justice* (1st ed., pp. 1–16). Routledge. https://doi.org/10.4324/9781003141457-1

Alvaredo, F., Chancel, L., Piketty, T., Saez, E., & Zucman, G. (2018). The elephant curve of global inequality and growth. *AEA Papers and Proceedings, 108*, 103–108. https://doi.org/10.1257/pandp.20181073

Ambrosini, M. (2020). *L'invasione immaginaria: L'immigrazione oltre i luoghi comuni (Prima edizione).* GLF editori Laterza.

Anderson, B. (2017). The politics of pests: Immigration and the invasive other. *Social Research: An International Quarterly, 84*(1), 7–28. https://doi.org/10.1353/sor.2017.0003

Anderson, B. R. O. (2016). *Imagined communities: Reflections on the crigin and spread of nationalism* (Revised ed.). Verso.

Anonymous. (2010). What Does Climate Justice Mean in Europe? *A Discussion Paper. Via Climate Justice Action Email List, 26 March.*

Appadurai, A. (1990). Disjuncture and difference in the global cultural economy. *Theory, Culture & Society, 7*(2–3), 295–310. https://doi.org/10.1177/026327690007002017

Appiah, A. (2006). *Cosmopolitanism: Ethics in a world of strangers* (1st ed). W.W. Norton & Co.

Arendt, H. (1966). *The origins of totalitarianism.* Penguin Classics.

Atapattu, S. A. (2018). *Human rights approaches to climate change: Challenges and opportunities (first issued in paperback).* Routledge, Taylor & Francis Group.

Balibar, É. (2002). *Politics and the other scene.* Verso.

Bauman, Z. (2013). *Globalization: The human consequences* (1. Aufl). Polity.

Beck, U. (2006). *The cosmopolitan vision.* Polity.

Beck, U., & Sznaider, N. (2006). A literature on cosmopolitanism: An overview. *The British Journal of Sociology, 57*(1), 153–164. https://doi.org/10.1111/j.1468-4446.2006.00098.x

Beckman, L., & Page, E. A. (2008). Perspectives on justice, democracy and global climate change. *Environmental Politics, 17*(4), 527–535. https://doi.org/10.1080/09644010802193393

Bhambra, G. K., & Newell, P. (2023). More than a metaphor: 'Climate colonialism' in perspective. *Global Social Challenges Journal, 2*(2), 179–187. https://doi.org/10.1332/EIEM6688

Calhoun, C. (2008). Cosmopolitanism and nationalism†. *Nations and Nationalism, 14*(3), 427–448. https://doi.org/10.1111/j.1469-8129.2008.00359.x

Carling, J., Carlino, J., & Sherpa, L. N. (2015). *Asia report on climate change and indigenous peoples*. Asia Indigenous Peoples Pact.

Celermajer, D., Schlosberg, D., Rickards, L., Stewart-Harawira, M., Thaler, M., Tschakert, P., Verlie, B., & Winter, C. (2021). Multispecies justice: Theories, challenges, and a research agenda for environmental politics. *Environmental Politics, 30*(1–2), 119–140. https://doi.org/10.1080/09644016.2020.1827608

Chouliaraki, L. (2013). *The ironic spectator: Solidarity in the age of post-humanitarianism*. Polity Press.

Chouliaraki, L., & Georgiou, M. (2022). *The digital border: Migration, technology, power*. New York University Press.

Climate Justice Now. (2010). *CJN! Network Members* (Climate-Justice-Now.Org/Category/Climate-Justice-Movement/Cjn-Members).

Comparelli, M. (2022). Intervista. In L. Tecleme (Ed.), *Guida rapida alla fine del mondo*. Tutto sulla crisi climatica e come risolverla.

Crenshaw, K. (1989). Demarginalizing the intersection of race and sex: A black feminist critique of antidiscrimination doctrine, feminist theory and antiracist politics. *University of Chicago Legal Forum, 1989*(8) https://chicagounbound.uchicago.edu/uclf/vol1989/iss1/8

Damon Matthews, H., Tokarska, K. B., Rogelj, J., Smith, C. J., Mac Dougall, A. H., Haustein, K., Mengis, N., Sippel, S., Forster, P. M., & Knutti, R. (2021). An integrated approach to quantifying uncertainties in the remaining carbon budget. *Communications Earth & Environment, 2*(1), 7. https://doi.org/10.1038/s43247-020-00064-9

Dastyari, A., Nethery, A., & Hirsch, A. L. (2023). *Refugee externalistation policies: Responsibility, legitimacy and accountability*. Routledge, Taylor & Francis Group. https://doi.org/10.4324/9781003167273

De Genova, N. (2013). Spectacles of migrant 'illegality': The scene of exclusion, the obscene of inclusion. *Ethnic and Racial Studies, 36*(7), 1180–1198. https://doi.org/10.1080/01419870.2013.783710

De Genova, N. (2017). *The borders of "Europe": Autonomy of migration, tactics of bordering*. Duke University Press.

De Genova, N. P. (2002). Migrant "illegality" and deportability in everyday life. *Annual Review of Anthropology, 31*(1), 419–447. https://doi.org/10.1146/annurev.anthro.31.040402.085432

de Sousa Santos, B. (Ed.). (2010). *Voices of the world*. Verso.

Derrida, J. (2000). HOSTIPITALITY. *Angelaki, 5*(3), 3–18. https://doi.org/10.1080/09697250020034706

Edenhofer, O., Wallacher, J., Lotze-Campen, H., Reder, M., Knopf, B., & Müller, J. (Eds.). (2012). *Climate change, justice and sustainability: Linking climate and development policy*. Springer.

Escobar, A. (2018). *Designs for the pluriverse: Radical interdependence, autonomy, and the making of worlds*. Duke University press.

Fanizza, F., & Omizzolo, M. (2018). *Caporalato: An authentic agromafia*. Mimesis International.

Fanon, F. (2001). *The wretched of the earth (reprinted)*. Penguin Books.

Federici, S. (2020). Calibano e la strega: Le donne, il corpo e l'accumulazione originaria. *Mimesis*.

Fraser, N. (Ed.). (2012). *Scales of justice: Reimagining political space in a globalizing world (Repr)*. Polity Press.

Gardiner, S. M. (2011). A perfect moral storm. *Oxford University Press*. https://doi.org/10.1093/acprof:oso/9780195379440.001.0001

Ghosh, A. (2016). *The great derangement: Climate change and the unthinkable (first published in Allen lane by penguin books India 2016)*. Allen Lane, an imprint of Penguin Books.

Giacomelli, E. (2023). *Panicocene: Narrazioni su cambiamenti climatici, regimi di mobilità e migrazioni ambientali*. Franco Angeli.

Giacomelli, E., & Walker, S. (2021). Challenging Eurocentric perceptions of mobility justice through climate diaries. *The Sociological Review Magazine. https://*. https://doi.org/10.51428/tsr.gxit8834

González, A. B. (2017). *Postcolonial approaches to Latin American children's literature*. Routledge.

Habtom, S., & Scribe, M. (2020). To breathe together: Co-conspirators for Decolonial futures. *Yellowhead Institute.*. yellowheadinstitute.org/2020/06/02/to-breathe-together

Han, H., & Ahn, S. W. (2020). Youth mobilization to stop global climate change: Narratives and impact. *Sustainability, 12*(10), 4127. https://doi.org/10.3390/su12104127

Haraway, D. J. (2016). *Staying with the Trouble: Making Kin in the Chthulucene* (p. dup;9780822373780/1). Duke University Press. doi:https://doi.org/10.1215/9780822373780.

Hardeman, R. R., & Karbeah, J. (2020). Examining racism in health services research: A disciplinary self-critique. *Health Services Research, 55*(S2), 777–780. https://doi.org/10.1111/1475-6773.13558

Harris, P. G., Chow, A. S. Y., & Karlsson, R. (2013). China and climate justice: Moving beyond statism. *International Environmental Agreements: Politics, Law and Economics, 13*(3), 291–305. https://doi.org/10.1007/s10784-012-9189-7

Hiskes, R. P. (2008). *The human right to a green future: Environmental rights and intergenerational justice* (1st ed.). Cambridge University Press. https://doi.org/10.1017/CBO9780511575396

Huntjens, P., & Zhang, T. (2016). *Climate justice: Equitable and inclusive governance of climate action (working paper 16)*. The Hague Institute for Global Justice.

IDMC. (2022). *Global Report on Internal Displacement 2022*. https://www.internal-displacement.org/global-report/grid2022/.

Imperatore, P., & Leonardi, E. (2023). *L'era della giustizia climatica*. Prospettive politiche per una transizione ecologica dal basso.

Indigenous Peoples' Global Summit on Climate Change. (2009). *The Anchorage Declaration*.

IPCC. (2022). *Climate Change 2022: Impacts, Adaptation, and Vulnerability* [Contribution of Working Group II to the Sixth Assessment Report of the Intergovernmental Panel on Climate Change].

Jackson, G., N'Guetta, A., De Rosa, S. P., Scown, M., Dorkenoo, K., Chaffin, B., & Boyd, E. (2023). An emerging governmentality of climate change loss and damage. *Progress in Environmental Geography, 2*(1–2), 33–57. https://doi.org/10.1177/27539687221148748

Jacques, P. J. (2006). The Rearguard of modernity: Environmental Skepticism as a struggle of citizenship. *Global Environmental Politics, 6*, 76–101.

Jamieson, D. (2010). *Ethics and the environment: An introduction (4. pr)*. Cambridge University Press.

Jones, R. (2017). *Violent borders: Refugees and the right to move (this paperback edition first published by verso 2017)*. Verso.

Jung, M., Dahal, P. R., Butchart, S. H. M., Donald, P. F., De Lamo, X., Lesiv, M., Kapos, V., Rondinini, C., & Visconti, P. (2020). A global map of terrestrial habitat types. *Scientific Data, 7*(1), 256. https://doi.org/10.1038/s41597-020-00599-8

Khosravi, S. (2011). *"Illegal" traveller: An auto-ethnography of borders (1. publ. in paperback)*. Palgrave Macmillan.

Kohn, E. (2013). How forests think: Toward an anthropology beyond the human. *University of California Press*. https://doi.org/10.1525/california/9780520276109.001.0001

Kronik, J., & Verner, D. (2010). *Indigenous peoples and climate change in Latin America and the Caribbean (English)* [Directions in development; environment and sustainable development]. World Bank. http://documents.worldbank.org/curated/en/654311468010837927.

Lama, P., Hamza, M., & Wester, M. (2021). Gendered dimensions of migration in relation to climate change. *Climate and Development, 13*(4), 326–336. https://doi.org/10.1080/17565529.2020.1772708

Latour, B. (1999). On recalling ant. *The Sociological Review, 47*(1_suppl), 15–25. https://doi.org/10.1111/j.1467-954X.1999.tb03480.x

Latour, B. (2017). *Facing Gaia: Eight lectures on the new climatic regime*. Polity.

Latour, B. (2018). *Down to earth: Politics in the new climatic regime*. (C. Porter, Trans.; (English ed.). Polity.

Lefebvre, H.(2000). La production de l'espace (4e éd). Anthropos.

Loyd, M. J., & Mountz, A. (2014). Managing migration, scaling sovereignty on islands. *Island Studies Journal, 9*(1), 23–42. https://doi.org/10.24043/isj.291

Malone, B. (2015). *Why Al Jazeera will not say Mediterranean 'migrants'* [Www.aljazeera.com]. Al Jazeera English, 20 August.

Mauro, E. (2018). *L'uomo bianco* (Prima edizione in "Serie bianca."). Feltrinelli.

Mbembe, A. (2019). *Necropolitics*. Duke University Press.

Mbembe, A. (2020). *Brutalisme*. La Découverte.

McNeill, J. R., & Engelke, P. (2016). The great acceleration: An environmental history of the Anthropocene since 1945. *The Belknap Press of Harvard University Press*. https://doi.org/10.4159/9780674970731

Meadows, D. H.with Club de Rome (Ed.) (1979). *The limits to growth: A report for the Club of Rome's project on the predicament of mankind* (2nd ed.). Universe books.

Meikle, M., Wilson, J., & Jafry, T. (2016). Climate justice: Between mammon and mother earth. *International Journal of Climate Change Strategies and Management, 8*(4), 488–504. https://doi.org/10.1108/IJCCSM-06-2015-0089

Mezzadra, S., & Neilson, B. (2013). *Border as method, or, the multiplication of labor*. Duke University Press.

Moellendorf, D. (2014). *The moral challenge of dangerous climate change: Values, poverty, and policy*. Cambridge University Press.

Mountz, A. (2011). The enforcement archipelago: Detention, haunting, and asylum on islands. *Political Geography, 30*(3), 118–128. https://doi.org/10.1016/j.polgeo.2011.01.005

Müller, M., & Walk, H. (2013). Demokratisierung der Klimaverhandlungssysteme durch verbesserte Partizipationsmöglichkeiten. In M. Dietz & H. Garrelts (Eds.), *Die internationale Klimabewegung* (pp. 57–74). Springer Fachmedien Wiesbaden. https://doi.org/10.1007/978-3-658-01970-9_3

Musarò, P., & Parmiggiani, P. (2022). *Ospitalità mediatica*. Le migrazioni nel discorso pubblico.

Neef, A., Pauli, N., & Salami, B. (Eds.). (2024). De Gruyter handbook of climate migration and climate mobility justice. *De Gruyter*. https://doi.org/10.1515/9783110752144

Newell, P. (2005). Race, class and the global politics of environmental inequality. *Global Environmental Politics, 5*(3), 70–94. https://doi.org/10.1162/1526380054794835

OHCHR. (2014). *Contribution of OHCHR's Indigenous Peoples and Minorities Section Contribution to the Thirteen session of the UN Permanent Forum on Indigenous Issues*. UN Office of the High Commissioner for Human Rights. https://www.un.org/esa/socdev/unpfii/documents/2014/ohchr.pdf

Papastergiadis, N. (2000). *The turbulence of migration: Globalization, deterritorialization, and hybridity*. Polity Press.

Papastergiadis, N. (2012). *Cosmopolitanism and culture*. Polity Press.

Pellizzoni, L. (2007). Il cambiamento climatico come oggetto culturale: Scienza, politica e incertezza. *Cosmopolis, 2*(1), 1–10.

Porter, L., Rickards, L., Verlie, B., Bosomworth, K., Moloney, S., Lay, B., Latham, B., Anguelovski, I., & Pellow, D. (2020). Climate justice in a climate changed world. *Planning Theory & Practice, 21*(2), 293–321. https://doi.org/10.1080/14649357.2020.1748959

Preciado, P. B. (2020). Learning from the virus. *Art*. https://www.artforum.com/print/202005/paul-b-preciado-82823

Rawls, J. (2005). *A theory of justice* (Original ed.). Harvard University Press. https://doi.org/10.2307/j.ctvjf9z6v

Robinson, S., & Carlson, D. (2021). A just alternative to litigation: Applying restorative justice to climate-related loss and damage. *Third World Quarterly, 42*(6), 1384–1395. https://doi.org/10.1080/01436597.2021.1877128

Sassen, S. (2007). A sociology of globalization. W. W. Norton.

Sassen, S. (2012). *Cities in a world economy* (4th ed.). SAGE/Pine Forge.

Saul, B., & McAdam, J. (2008). An insecure climate for human security? Climate-induced displacement and international law. In A. Edwards & C. Fernstman (Eds.), *Human security and non-citizens*. Cambridge University Press.

Schapendonk, J., Bolay, M., & Dahinden, J. (2021). The conceptual limits of the 'migration journey'. De-exceptionalising mobility in the context of west African trajectories. *Journal of Ethnic and Migration Studies, 47*(14), 3243–3259. https://doi.org/10.1080/1369183X.2020.1804191

Scheel, S., & Tazzioli, M. (2022). Who is a migrant? Abandoning the nation-state point of view in the study of migration. *Migration Politics, 1*(1), 002. https://doi.org/10.21468/MigPol.1.1.002

Schlosberg, D., & Collins, L. B. (2014). From environmental to climate justice: Climate change and the discourse of environmental justice. *WIREs Climate Change, 5*(3), 359–374. https://doi.org/10.1002/wcc.275

Sen, A. (2009). *The idea of justice*. Penguin Books.

Sheller, M. (2018). *Mobility justice: The politics of movement in the age of extremes*. Verso.

Sheller, M. (2020a). Island futures: Caribbean survival in the Anthropocene. *Duke University Press*. https://doi.org/10.1215/9781478012733

Sheller, M. (2020b). Mobility justice in urban studies. In O. B. Jensen, C. Lassen, V. Kaufmann, M. Freudendal-Pedersen, & I. S. Gøtzsche Lange (Eds.), *Handbook of urban Mobilities* (1st ed.). Routledge. 10.4324/9781351058759.

Sheller, M., & Urry, J. (2006). The new Mobilities paradigm. *Environment and Planning A: Economy and Space, 38*(2), 207–226. https://doi.org/10.1068/a37268

Shilliam, R. (2015). *The black Pacific: Anti-colonial struggles and oceanic connections*. Bloomsbury Academic.

Shue, H. (2016). *Climate justice: Vulnerability and protection (first published in paperback)*. Oxford University Press.

Smith, P. (2011). Narrating global warming. In J. C. Alexander, R. N. Jacobs, & P. Smith (Eds.), *The Oxford handbook of cultural sociology*. Oxford University Press.

Squire, V. (Ed.). (2012). *The contested politics of mobility: Borderzones and irregularity (first issued in paperback)*. Routledge Taylor & Francis Group.

Srinivasan, K., & Kasturirangan, R. (2016). Political ecology, development, and human exceptionalism. *Geoforum, 75*, 125–128. https://doi.org/10.1016/j.geoforum.2016.07.011

Sullivan, P. (2009). *Lift every voice: The NAACP and the making of the civil rights movement*. New Press.

Sultana, F. (2022a). Critical climate justice. *The Geographical Journal, 188*(1), 118–124. https://doi.org/10.1111/geoj.12417

Sultana, F. (2022b). The unbearable heaviness of climate coloniality. *Political Geography, 99*, 102638. https://doi.org/10.1016/j.polgeo.2022.102638

Sultana, F. (2024). Confronting climate Coloniality: Decolonizing pathways for climate justice (1st ed.). Routledge. doi:https://doi.org/10.4324/9781003465973.

Torpey, J. C. (2018). *The invention of the passport: Surveillance, citizenship and the state (second edition)*. Cambridge University Press.

UNHCR. (2022). Global trends report 2022. UN Office of the High Commissioner for Refugees. https://www.unhcr.org/global-trends-report-2022

Urry, J. (2002). The tourist gaze (2. ed.). Sage.

Verlie, B., & Neimanis, A. (2023). Breathing climate crises: *Feminist environmental humanities and more-than-human witnessing. Angelaki, 28*(4), 117–131. https://doi.org/10.1080/0969725X.2023.2233810

Villani, L. (2025). *Essere umano, essere terrestre: Immaginari e pratiche di giustizia climatica in Ultima Generazione*. [PhD dissertation]. University of Bologna.

Visvanathan, S. (2007). 6. Knowledge, justice and democracy. In M. Leach, I. Scoones, & B. Wynne (Eds.), *Science and citizens: Globalization and the challenge of engagement*. Zed books.

Viveiros de Castro, E. (2015). *The relative native: Essays on indigenous conceptual worlds*. Hau Books.

Walker, S., & Giacomelli, E. (2024). 16 no country for young women and men? Youth perspectives on the climate crisis and (Im)mobility in Senegal. In A. Neef, N. Pauli, & B. Salami (Eds.), *De Gruyter handbook of climate migration and climate mobility justice* (pp. 338–361). De Gruyter. https://doi.org/10.1515/9783110752144

Whyte, K. (2016). Indigenous environmental movements and the function of governance institutions. In T. Gabrielson, C. Hall, J. Meyer, & D. Schlosberg (Eds.), The Oxford handbook of environmental political theory (pp. 563–580). Oxford university press.

Wiegel, H., Boas, I., & Warner, J. (2019). A mobilities perspective on migration in the context of environmental change. *WIREs Climate Change, 10*(6), e610. https://doi.org/10.1002/wcc.610

Wihtol de Wenden, C. (2015). Il diritto di migrare. *Ediesse*.

Wilson, T. M., & Donnan, H. (Eds.). (2000). *Border identities: Nation and state at international frontiers (reprint)*. Cambridge Univ. Press.

Wimmer, A., & Schiller, N. G. (2003). Methodological nationalism, the social sciences, and the study of migration: An essay in historical epistemology. *International Migration Review, 37*(3), 576–610. https://doi.org/10.1111/j.1747-7379.2003.tb00151.x

WTO. (2023). *WTO annual report 2023*. World Trade Organization. https://www.wto.org/english/res_e/publications_e/anrep23_e.htm

Zanfrini, L. (2016). *Introduzione alla sociologia delle migrazioni (Prima edizione)*. GLF editori Laterza.

CHAPTER 3

Narratives of Climate Change and Migration

Introduction

In its most recent report, the Intergovernmental Panel on Climate Change (IPCC), provides a comprehensive overview of the scientific evidence demonstrating that human activities are driving unprecedented—and, in some cases, irreversible—changes to the global climate (IPCC, 2022). The report emphasizes that climate change is predominantly anthropogenic and highlights the potential for mitigation through deliberate interventions aimed at halting, decelerating, or transforming current trajectories. It also underscores growing public concern with the tangible impacts of climate change, as evidenced by questions arising during extreme events such as rising summer temperatures, hurricanes along the US coastline, floods across European river systems, and the opening of new shipping routes as a result of polar ice cap retreat. Yet such phenomena are often met with public and media inquiries about whether they can definitively be linked to climate change.

Why, then, is there such a persistent gap between the abundance of scientific information and the amount of public awareness or concern? Could it stem from scientists' inability to effectively communicate their findings to the public? Is It the result of the media's failure to report these findings clearly and accurately? Or should we instead focus on policymakers who ignore or downplay the warnings issued by the scientific community?

E. Giacomelli and P. Musarò, *Climate Mobility Justice*,
https://doi.org/10.1007/978-3-032-07907-7_3

Similarly, when it comes to migration, numerous researchers have observed a disconnect between the resident population's perception of immigration levels and the real number of immigrants recorded in official statistics. In recent decades, immigration has often been at the center of the political agenda. In Europe and other so-called Global North countries, politicians and the media have mainly obsessed over irregularized immigration, security, and crime, sidelining issues such as integration and the structural management of migrants themselves. The result is a widespread misperception: citizens overestimate both the scale and the supposed threat of migration, reinforcing a cycle of fear and suspicion.

In Italy, the dominant media narrative presents immigration as incessant, chaotic, and threatening—largely composed of young men, mainly refugees and irregular immigrants, mostly from Africa and the Middle East, often Muslim (and therefore suspected of terrorism), portrayed as stealing work from Italians and burdening the Italian welfare system. This narrative of an influx of people invading our territories ends up consolidating the idea of a perennial 'migratory crisis or emergency'—a tale that does not correspond to reality: there is, in fact, a lower percentage of the (regular) immigrant population compared to the total population (between 7 and 8% in European countries), and an even lower percentage of refugees (less than 3%).[1]

What this disjunction between expert knowledge and public perception reveals is the urgent need to interrogate how knowledge is communicated, contested, and co-opted in public discourse. Why is scientific data—accessible, intelligible, and well-documented—so easily overshadowed by populist rhetoric and sensationalist framing? Data on citizens' perceptions prompt critical reflection on the factors influencing individual viewpoints, particularly in the formation of public opinion, which is intricately tied to media representation and political discourse on the issue. Through the lens of justice, this chapter critically examines the complex, interrelated dynamics between climate change and various

[1] According to UNHCR data, by the end of 2023, the number of people displaced across borders hosted in European countries was 13 million. This is an increase of about 600,000 compared to 2022, primarily resulting from the full-scale war on Ukraine, which started in February 2022 and has resulted in one of the largest human displacement crises globally and the largest refugee influx in Europe in decades. Cfr. https://www.unhcr.org/about-unhcr/where-we-work/europe.

mobility regimes by addressing the dissonance between reality and perception. To this end, we explore how different strategies have been employed to either internalize or externalize knowledge, undermine experts, or exclude them from public discourse—sometimes even denigrating or outright denying their contributions and credibility.

The very notion of communication, from the Latin *communis* (meaning 'belonging to all'), reminds us that to communicate is to share or make something in common among individuals. As Raymond Williams argued in *Keywords* (1976), communication is not simply about conveying information, but about shaping and negotiating shared social meanings. Analyzing communication therefore requires us to interrogate not just content, but also the codes, contexts, and power relations that structure how messages are encoded, decoded, and made intelligible (Artieri et al., 2022).

How we communicate is seldom neutral. The language we use conveys differences in the representation of the world and plays an important role in the (re)production, (re)creation, and transformation of social imaginaries, and in the social construction of reality (Berger & Luckmann, 1990). Inaccurate or ideologically driven language doesn't merely distort our understanding—it affects our emotions, decisions, and collective responses.

This chapter thus examines how certain terms and rhetorical framings frequently associated with climate change and migration shape our understanding—sometimes illuminating complexity, but more often obscuring it. It interrogates the role of dominant narratives in naturalizing climate crises and depoliticizing the discourse on migration.

Drawing on Jacques Rancière's concept of the 'distribution of the sensible' (Rancière, 2009) we investigate how political and media discourses delineate what is visible, audible, and sayable—legitimizing some experiences and epistemologies while occluding others. In the process, this chapter addresses the framing tools that structure public discourse and influence affective responses to both climate and migration. It seeks to uncover both the cognitive traps that individuals—whether citizens, audiences, media practitioners, or policymakers—tend to fall into when engaging with these phenomena. It also tackles the underlying power dynamics embedded within certain narratives and probes their implications for shaping effective responses. By challenging established perceptions, we aim to advocate for more nuanced communication strategies that foster intersectional understanding, deconstruct entrenched assumptions of agency, reframe fear-based narratives, expose harmful tendencies, and dismantle common misconceptions. Such strategies are essential for constructing a public sphere in which collective imagination is anchored in positive and empowering reference points.

More specifically, the first part of the chapter de-naturalizes climate change by highlighting its social, cultural, and economic dimensions, and rejecting the trap of its 'supposed naturalness.' Indeed, it is important to deconstruct the concept of 'environment' in the narrative of the climate crisis and shift the emphasis onto those processes that have fostered its use and/or abuse. We critique how 'the environment' is constructed in dominant climate narratives, especially through anthropocentric, racially-blind (Yusoff, 2018), and gender-blind (Xausa, 2020). Narratives of natural disasters frequently oscillate between vague uncertainty and sensationalized techno-utopian or dystopian-melodramatic prophecies of impending apocalypses—often leaving the public to wallow in fatalistic resignation (Giacomelli, 2023).

The second section turns to mobility and migration, engaging the conceptual apparatus of Mimi Sheller's *Mobility Justice* (Sheller, 2018). Since 2011, migration in Europe has been persistently framed as a 'crisis,' a term that simplifies complex social processes and justifies increasingly restrictive policies.

To liberate the study of migration from the conceptual constraints it has been ensnared in, this section examines how alarmist storytelling has come to dominate public discourse over the past three decades. Drawing on Rancière's concept of the 'distribution of the sensible,' this approach elucidates how the toxic narratives surrounding climate change and migration are neither natural nor inevitable but are actively constructed to uphold certain hierarchies and exclusions. By challenging and reshaping this distribution—by altering what is visible, audible, and articulated—it becomes possible to reveal the exclusionary and harmful nature of these narratives and to advance toward a more equitable and inclusive understanding of these interconnected global issues.

The Social Construction of Reality

When reflecting on the role of media and narratives in shaping our perception of either migration or the climate crisis, it is crucial to refer to the concept of 'imaginary,' which indicates the symbolic repertoire that a social system uses to communicate meaning (Taylor, 2003; Chouliaraki, 2013). In his seminal work *The Imaginary Institution of Society* (1987), Castoriadis explored how institutions (language, the economy, and government) are products of the social imaginary: the shared frameworks, representations, and meaning-making practices that underpin a society's norms, organizing bodies, and values. These imaginaries are not tangible objects or simply reducible to material or economic factors. Rather, they are the collective creations that give a society coherence, providing it with its unique identity. Nations, democracy, capitalism, or religion can all be seen as products of the social imaginary—they are systems that exist because people collectively imagine and uphold them as meaningful. As living forces, imaginaries shape human existence—yet they are also contested and reconfigurable. Through more radical imaginaries, individuals and societies can challenge inherited constructions and create new possibilities.

In Rancière's terms, it is through the 'distribution of the sensible' that power structures determine what is visible, sayable, and knowable within a given order. This regime of perception defines what counts as real, what remains excluded, and who has the authority to speak. Narratives operate within this logic: they legitimize certain positions as the only viable or intelligible perspectives. For instance, when the media systematically omit the voices of marginalized communities, it reinforces the idea that these communities have nothing important to say, thereby consolidating their invisibility in the public sphere.

Narratives are instruments for organizing human cognition (Boland & Tenkasi, 1995), which can be defined as 'discourses with a clear sequential order that connect events in a meaningful way for a definite audience and thus offer insights about the world and/or people's experiences of it' (Hinchman & Hinchman, 1997, p. XVI). As Beyers notes, narratives can be interrogated through three key features: 'First, they are *temporal* in that they represent events as part of sequences which have a beginning, a middle, and an end. Second, they are *meaningful*. One important way in which they convey meaning is by ordering events into a temporal sequence that relates them to a definite resolution or conclusion. This is why narratives can be said to have a causal dimension. Thirdly, they are inherently *social* as they are produced for specific audiences. It follows that, they are only meaningful for those audiences,' though, naturally, unintended recipients may happen on them (2006, p. 2).

As we will see in the following pages, narratives operate performatively—that is, through familiar aesthetic practices, they solicit affective engagement and shape how we interpret and act in the world in line with what is deemed desirable by a specific culture. This performative power is central to how certain visions of justice, suffering, or crisis come to be articulated, contested, and institutionalized over time. This focus on performance as a conduit of inquiry into the social imaginary of climate change and migration invites us to consider two related dimensions of the imaginary: (1) its historicity and (2) its performativity. The first refers to how imaginaries evolve over time, generating new claims about causes, effects, and responsibilities. The second refers to how these shifting imaginaries produce new truths about suffering and social justice—within and beyond Western frameworks.

In this sense, the act of migrating is inseparable from the migratory imaginary of the last three decades, shaped both by the North's gaze of fear and anxiety and the South's gaze of aspiration. To grasp the reality of migration, we must investigate not only the material conditions of movement but also the narratives that precede, accompany, and follow it—narratives constructed through media frames that filter meaning about the genesis, processes, and consequences of migration.

Similarly, climate change is often framed as a 'natural' challenge, or anthropogenic yet resolvable through market-driven mechanisms or technological innovation. This framing overlooks how the carbon economy's political structures and the enduring myth of infinite growth have historically shaped societal imaginaries of possible futures, perpetuating these narratives as foundational to modern progress. Denaturalizing climate change is thus imperative if we want to critically examine its embedded social, economic, political, and ethical interconnections. Such an analysis exposes the power dynamics and discursive obfuscations latent in the terminology used to talk about the crisis and challenges the narrative mystification of its systemic roots.

More broadly, insofar as public opinion is formed through media narratives, the latter's *agenda-setting* (McCombs, 2005) and *framing* (Entman, 1993; Goffman, 1986) plays an important role in influencing political attitudes and steering public debates on 'otherness' (Bourdieu, 1980), climate change, migration, and asylum. According to Goffman (1986), *framing* refers to the selective emphasis placed on certain aspects of a situation, issue, or story while downplaying or omitting others. Frames provide a structured lens through which people make sense of complex realities. Thus, narratives embedded within frames can delimit what seems thinkable or actionable, foreclosing some political options while making others appear more reasonable, or even inevitable. Narrative then becomes a tool through which to influence others, bringing them closer to one's view of things by giving greater prominence to certain story elements over others.

Framing works through the activation of mental schemas—pre-existing knowledge structures or beliefs that shape how individuals process information. By choosing frames that align with these established schemas, communicators can evoke specific emotional and cognitive responses. According to this theory, the frame, in addition to defining the problem, also indicates its causes and proposes solutions, generally implying a choice of direction in terms of policy or implicit answers to questions such as 'what should be done' about an issue. Further, the framing effect can be considered as an extension of agenda-setting in the domain of interpretation and mediated meaning-making. In McCombs' own words, 'the media are not only successful in telling us what to think, but they are also successful in telling us how to think about such issues' (McCombs, 2005, p. 546). In choosing certain aspects of perceived reality, they make them more 'salient' to promote 'a particular problem definition, cause interpretation, moral evaluation, and/or recommendation' (Entman, 1993, p. 52).

Conflicting Imaginaries of Climate (in)Justice

From an epistemological standpoint, the media play a pivotal role in constructing crises: they name, frame, and assign meaning to them. Thus, it is essential to view climate change narratives as forms of knowledge that articulate compelling imaginaries through representations of past and present scenarios, as well as projections of future possibilities (Krzyżanowski & Forchtner, 2016). For media scholars, this calls for a double analytical movement. First, we must trace how the growing conceptualization of these issues in public discourse can often obscure the actions and agents underpinning these concepts, thereby hindering the identification of intricate networks of beneficiaries, victims, and other figures of accountability. Second, it demands critical scrutiny of the specific concepts being coined, disseminated, endorsed, or contested within contemporary media-saturated environments (Kunelius & Roosvall, 2021).

What are the implications of analyzing climate change and mobility regimes through the lens of justice? To answer this, we must identify the cognitive and communicative barriers that prevent societies from grasping the true nature of the crisis, finding consensus on a shared conception of justice, and acting accordingly. These barriers are rooted in the same dynamics that create a disconnect between expert knowledge and populist rhetoric—rhetoric that shapes public opinion by legitimizing specific 'spheres of the sensible' (Rancière, 2009) while concealing power relations. Delegitimization, silencing, and erasure are central strategies in this process, and they define which solutions are imagined and pursued.

How societies imagine their climate futures will significantly influence how those futures materialize. This raises urgent questions: How can we uncover, understand, elevate, engage with, nurture, communicate, and implement diverse (future) climate imaginaries that foster equitable and effective responses to a rapidly changing climate? Building on the work of Davoudi and Machen (2022), climate imaginaries may be conceptualized as 'socio-technical practices, culturally situated, that create collective meaning about the past, present, and future. [These imaginaries] are co-constitutive of the material and discursive means through which they are produced, circulated, normalized, and contested by political actors' (p. 207). In other words, the interplay between cultural narratives and political agency deeply shapes how societies imagine and act upon climate futures.

The climate issue poses several 'traps' and risks for future climate imaginaries, some of which are crucial to human civilization (Silvestrini & Zorzoli, 2020). Among the various challenges, two of the most significant and closely interrelated obstacles are the *cognitive* trap and the *communicative* trap. The first stems from the failure to fully grasp the urgency and gravity of the crisis, which in turn fails to prompt the necessary actions. As an editorial in *The Economist* (September 23, 2019) put it, global warming is pictured as 'a diabolical problem for humanity—urgent yet addressed in slow motion, immediate yet perceived as distant, real yet simultaneously abstract.'[2] The climate crisis exhibits an epistemological disjuncture: public risk perception and the necessary implementation of mitigation strategies remain misaligned with the phenomenon's existential gravity. This disjunction arises partly from climate change's status

[2] https://www.economist.com/leaders/2019/09/19/the-climate-issue.

as a hyperobject—a conceptual framework advanced by Timothy Morton (2012) to describe entities whose vast spatial and temporal scales evade direct human apprehension. As Greenpeace director Giuseppe Onufrio observes, the 'biblical'[3] magnitude of projected climatic disruptions triggers paradoxical psychological mechanisms—ranging from cognitive dissonance and message dissociation to fatalistic resignation—that collectively immobilize transformative collective action. These responses stem from the tension between the crisis's planetary immediacy and its perceptual abstraction, creating a feedback loop where delayed recognition perpetuates inadequate political and societal responses (Armiero & De Angelis, 2017).

Closely linked is the communicative trap, which concerns how narratives about climate change are (or are not) made desirable, actionable, and imaginable. For George Monbiot, the media (television, newspapers, radio, magazines, etc.) are 'the industry [most] responsible for the destruction of life on Earth' (Monbiot, 2022, p. 369), since it sustains the ideological consensus on which our Earth-devastating economic and productive system depends on. Chakrabarty (2017) reminds us that 'if we do not understand modernity as an object of desire, we cannot understand why it remains seemingly viable despite its many obvious problems.' Alex Langer (1994) puts it more succinctly when he asserts that 'ecological conversion will only be able to assert itself if it appears socially desirable.' Thus, the communicative trap refers to the ways in which a radical shift in addressing climate change is (or is not) articulated, represented, and rendered both desirable and conceivable.

Shapiro (1988) points out that representations do not imitate reality; rather, they are practices through which things take on meaning and, in our case, force us 'to live [...] in many different ends of the world' (Malvestio, 2021, p. 13). It is from this perspective that Ghosh (2016) argues that an imaginative and cultural failure lies at the heart of the climate crisis. But what are, more precisely, the dominant climate change

[3] http://www.theguardian.com/environment/2023/feb/14/rising-seas-threaten-mass-exodus-on-abiblical-scale-un-chief-warns.

imaginaries? What (toxic) narratives[4] are told—and by whom—about the 'world's biggest story' (Thunberg, 2022, p. 41)?

The narratives surrounding climate change are often presented with varying degrees of certainty. At times, they are imbued with techno-utopian optimism or dystopian melodrama, featuring apocalyptic forecasts and fatalism. This analysis seeks to examine how media framing—through the selective amplification of voices such as scientists predicting outcomes, government officials managing responses, and survivors recounting their experiences—shapes the discourse on natural disasters. By determining which perspectives are prioritized, the media effectively influences who is granted authority to define and steer public opinion on these catastrophic events.

Denialist Blindness

In 2011, shortly after taking office, Florida's Republican governor Rick Scott implemented an unofficial ban on terms such as 'climate change,' 'global warming,' 'sustainability,' and 'sea-level rise' in official government communications. Scott justified this decision by asserting that these terms would only incite unwarranted alarmism, further claiming that climate change was 'not even caused by human activity.'[5] Ironically, Florida—with its extensive low-lying coastline, vulnerability to late-summer hurricanes, and susceptibility to rising sea levels—is among the US states most exposed to the risks of flooding and other extreme weather events—phenomena widely linked by scientists to global warming. The destructive impacts of recent hurricanes such as Irma (2017) and Ian (2022) underscore this vulnerability. Despite the absence of a formal enforcement protocol, employees of Florida's Department of Environmental Protection complied with the directive by limiting their language to terms deemed factual, effectively aligning themselves with the ideology of climate change denialism prevalent within the Republican neo-right.

[4] Our sense of toxic narratives echoes the one of the Wu Ming collective (2013), who write "to become toxic narrative, a story must always be told from the same point of view, in the same way and with the same words, always omitting the same details, removing the same elements of context and complexity." As Marco Armiero (2021) writes, toxic narratives are powerful and dangerous because they domesticate collective memories and vanquish alternative practices of remembrance.

[5] http://fcir.org/2015/03/08/in-florida-officials-ban-term-climate-change/.

This approach reflects a broader denialist strategy: 'if the word does not exist, neither does the problem' (Fargione & Concilio, 2018, p. 15).

In the early hours of his return to the White House in January 2025, President Donald Trump signed an executive order titled *Restoring Freedom of Speech and Ending Federal Censorship*. While ostensibly aimed at safeguarding free speech, the order quickly revealed its underlying intent: imposing strict controls over language and redefining key terms across federal agencies. Among other directives,[6] the administration ordered the redefinition of 'climate change'[7] and excluded solar and wind power from the official definition of 'energy.' Within days, references to climate change and related concepts began disappearing from federal websites, with similar actions taken by other institutions. Scientists[8] reliant on federal funding were instructed to halt any activities that contradicted Trump's mandates, while government employees resorted to convert measures to continue climate-related work, all the while fearing that even acknowledging global warming could jeopardize their positions.[9] This suppression of terminology reflects the emergence of a reactionary movement known as the 'woke right,' which enforces its own cultural lexicon in opposition to progressive discourse. The removal of climate-related language from over 200 government websites, as documented by the Environmental Data and Governance Initiative (EDGI), illustrates how such policies shape public understanding by obscuring scientific realities.[10] This erasure deepens ideological divides, making it increasingly difficult for society to address pressing environmental challenges. As Gretchen Gehrke of EDGI has noted, the removal of authoritative information undermines collective recognition of reality, leaving individuals vulnerable to fragmented perceptions shaped by social media rather than reliable sources. The consequences of these actions extend beyond

[6] Especially on diversity, equity, and inclusion programs.

[7] https://www.theguardian.com/environment/2025/feb/21/trump-scientific-research-climate.

[8] https://edition.cnn.com/2025/02/21/climate/trump-blocks-scientists-ipcc/index.html.

[9] https://theconversation.com/revoking-epas-endangerment-finding-the-keystone-of-us-climate-policies-isnt-simple-and-could-have-unintended-consequences-252271.

[10] https://sunlightfoundation.com/2019/07/22/environmental-data-and-governance-initiative-edgi-releases-new-report-detailing-the-extent-of-censorship-of-climate-change-related-terms-on-federal-websites/.

rhetoric: federal agencies have faced mass layoffs, funding freezes, and the dismantling of environmental initiatives, further weakening institutional capacity to address climate change. Scientists and researchers have described this new climate as demoralizing, with many struggling to continue their work under such restrictive conditions. The administration's aggressive stance on language and policy not only hinders efforts to combat global warming but also sets a precedent for how political agendas can manipulate public discourse and scientific integrity.

In *The great derangement: Climate change and the unthinkable* (2016), Amitav Ghosh examines the predominance of global warming research originating from the Anglosphere.[11] He contends that this region exhibits a distinct cultural and geographic blindness toward climate change, shaped by decades of narratives that either dismiss or minimize its significance. Ghosh attributes this phenomenon to several interconnected factors: colonial and imperial legacies that prioritize exploitation over sustainability, an ideological focus on economic growth coupled with technological optimism as panaceas, and a persistent refusal to truly acknowledge the global scope of climate change.

Narratives that undermine or deny global warming and its anthropogenic origins have profoundly influenced both public policy and societal perceptions. A historical example is 'the big climate lie,' which emerged in the United States in the fall of 1992, when E. Bruce Harrison—a public relations specialist often regarded as the pioneer of climate denialism—asserted that the greenhouse effect was no longer an established scientific reality but rather an uncertain phenomenon with unpredictable consequences. Harrison's approach involved reframing climate discourse; he secured an annual contract valued at $500,000 (approximately €960,500 today) from the Global Climate Coalition (GCC). This coalition represented industries such as oil, coal, automotive manufacturing, utilities, steel production, and railroads—all seeking strategic communication services to reshape public understanding of climate change narratives (BBC News).[12]

In *Climate Liars: Power, Politics, and the Psychology of Century Crisis Deniers* (2021), Stella Levantesi investigates what she describes as the

[11] The term refers to all English-speaking cultures and societies, including English-speaking nations and their diasporas.

[12] www.bbc.com/news/science-environment-62225696.

most significant cover-up operation of the century: the deliberate denial of the climate emergency. For Levantesi, one key factor behind the rise of climate denialism is the inherent 'invisibilities' of planetary changes, which are often imperceptible in real time. This lack of immediate perception, she argues, is exploited to mislead public opinion about the true causes of climate change (Fargione & Concilio, 2018, p. 17). Furthermore, climate deniers have historically received equal or even greater media attention than climate scientists—the more their narratives were amplified, the more scientific consensus became undermined (Monbiot, 2022).

The Anglosphere provides numerous examples of denialist narratives that have shaped public discourse. For instance, Bjorn Lomborg's *The Skeptical Environmentalist* (Lomborg, 2001) downplays the severity of global warming, suggesting that it is exaggerated by the media and that proposed solutions are both costly and ineffective. Similarly, the British documentary *The Truth About Climate Change* (2006) controversially frames the fossil fuel industry not as a primary cause of global warming but as a potential solution through their use of technologies such as carbon capture and storage. Another example is the film *The Great Global Warming Swindle* (2007), which attributes global warming primarily to natural phenomena like solar activity rather than human-induced greenhouse gas emissions.

Looking beyond individual cultural artefacts' perpetuation of climate denialism, it is essential to recognize that all cultures exhibit a degree of self-blindness. As José Saramago reflects in his novel *Blindness* (Saramago, 2021), 'We have not become blind; in my opinion, we are blind people who see, blind people who, while seeing, do not see.' This blindness does not obscure awareness of climate change itself—at least not anymore—but rather renders the future unimaginable because 'the present is already charred even in its imagination' (Cappi, 2023, p. 19). It is this imaginative exhaustion that leads to the well-known maxim, popularized by Mark Fisher in *Capitalist Realism* (Fisher, 2009) and often attributed to Slavoj Žižek or Fredric Jameson: 'It is easier to imagine the end of the world than the end of capitalism.'

Techno-Utopian Narratives

Techno-utopian narratives of climate change, which often align with the Anthropocene's emphasis on human centrality, express an optimistic belief in the capacity of technology and human innovation to resolve environmental challenges (Ribeiro & Soromenho-Marques, 2022). These narratives are frequently positioned as hopeful, utopian alternatives to apocalyptic accounts of a dystopian future. They typically advocate for a combination of technological, scientific, and engineering advancements that can effectively address climate change and ensure the planet's sustainability for future generations. However, this approach—characterized by its blindness to race and gender, and its predominantly anthropocentric outlook—pervades many dominant discourses surrounding the climate crisis (Xausa, 2020). As Andrew Baldwin (2017) observes, the archetypal figure in such narratives is often a white, European, cisgender, Western scientist.

These techno-utopian narratives often frame climate change primarily as a scientific—or 'scientificalized'—problem, emphasizing technological and economic solutions while frequently overlooking the finite nature of natural resources and the imperative of long-term environmental sustainability (MacGregor, 2009). This prompts a critical question: are these optimistic portrayals of climate change merely an example of 'technowashing'? Borrowed from the notion of greenwashing[13]—defined as actions or behaviors that create the false impression that a company is doing more for environmental protection than it actually is—technowashing refers to narratives that exhibit excessive confidence in techno-scientific solutions as the sole remedy for environmental challenges. Simultaneously, such narratives often promote grandiose technological visions that divert attention from the root causes of ecological problems (Ribeiro & Soromenho-Marques, 2022, p. 5).

Gerry Canavan (2021) describes the present era as 'sci-fi times,' referencing the prevalence of media narratives that deploy science and technology to conjure (in)desirable imaginaries and oversimplified solutions to climate change. These techno-utopian frameworks raise several pressing concerns. First, they often divert attention from the necessary and urgent critique of the neoliberal economic model underpinning the world's largest and most ecologically destructive economies. Second, they reinforce the idea of human dominance over nature, perpetuating the belief that humanity can overcome each ecological crisis with ever-newer technologies. Third, they serve to legitimize unregulated markets by

[13] Cambridge Dictionary. Greenwashing: https://dictionary.cambridge.org/dictionary/english/greenwashing.

preserving corporate reputations and ensuring uninterrupted economic growth. Investments in such technologies frequently deflect scrutiny from the environmentally harmful practices of both states and corporations. Finally, these narratives shift accountability away from political leaders, positioning science and technology—not governance or systemic reform—as the key drivers of future action (Ribeiro & Soromenho-Marques, 2022).

Techno-utopian narratives about climate change are frequently echoed in the rhetoric of politicians and affluent investors. Al Gore's documentary *An Inconvenient Sequel: Truth to Power* (2017), for example, underscores the potential of technological solutions such as solar and wind energy, while showcasing ongoing positive transformations in the adoption of renewables. Similarly, Elon Musk's SpaceX Mars program exemplifies this narrative by envisioning human colonization of Mars as early as 2026, with aspirations for settlement and space tourism in the decades to follow. Musk's vision of a self-sustaining Martian colony reflects a techno-utopian faith in technology as a solution to existential threats and a means of expanding humanity's planetary frontier. In such a narrative, heterosexual men from affluent, globalized regions often cast themselves as visionary problem-solvers, actively engaging the public imagination through their investments in science and innovation. Yet their proposals frequently overlook critical questions of feasibility and justice, while underestimating the complex social, economic, and political dimensions of the climate crisis.

The Apocalyptic Narratives

In recent years, the heightened media focus on the climate 'emergency' has amplified the role of crisis narratives as perceived 'tipping points.' The declaration of a 'climate emergency' in 2019[14] marked a significant shift, transforming climate change into a media-driven and mediatized 'crisis,' modeled after archetypes taken from the apocalyptic imaginary (Schuster, 2014). This framing has contributed to the mobilization of youth-led movements such as Fridays for Future, Extinction Rebellion, and Last Generation, which demand accountability and urgent action. Yet the alarmist portrayal of climate change as a looming apocalypse has

[14] https://climateemergencydeclaration.org/climate-emergency-declarations-cover-15-million-citizens/.

also fostered distorted and dystopian perceptions of the crisis—at times leading to paralysis and crippling inaction.

As Lawrence Buell (1996) notes, the apocalypse represents one of the most powerful metaphors available to the ecological imagination. Still, it is crucial to interrogate the implications of the frequent (ab)use of crises and end-of-world narratives. Given that emergencies are by definition temporary phenomena, applying the term to a long-term structural process like the climate crisis raises significant concerns (Abbas, 2012).

As previously discussed, effectively addressing climate change requires long-term transformations in societal organization, economic models, and lifestyle practices. Framing climate change as an emergency, rather than as a systemic byproduct of the capitalist growth imperative, carries several significant drawbacks. Hodder and Martin (2009) contend that apocalyptic framings of climate change can be counterproductive—they risk overwhelming the public, presenting the crisis as insurmountable—a 'hyperobject' fostering despair, rather than galvanizing structural change. This narrative frame is not a novel phenomenon. In his 1967 analysis of apocalyptic narratives, literary scholar Frank Kermode argued that 'the concept of crisis is a central element in the way we make sense of reality,' arguing that people often interpret their own time as exceptionally close to the end. According to Kermode, 'time is not free; it is subject to a mythical end,' which leads to a tendency to perceive current crises as uniquely threatening (Kermode et al., 2020, p. 112). Similarly, in *The End of the World: A Contribution to the Analysis of Cultural Apocalypses* (2023), Italian anthropologist Ernesto de Martino asked the question 'Where do apocalypses go?', exploring how end-of-the-world fantasies can result in cultural and psychological stagnation, manifesting either as denialism or delusions of total destruction.

A growing body of research (Nixon, 2013; Yusoff & Gabrys, 2011; Duffy & Leppänen, 2024; Schinaia, 2025) has shown that climate change is most often framed through the lens of natural disasters and apocalyptic catastrophes. While this frame captures media attention, it tends to overshadow more gradual yet devastating processes of 'slow violence'—the incremental and often invisible forms of environmental degradation and exploitation that disproportionately affect marginalized communities in the Global South. These slow and structural forms of violence, deeply rooted in neocolonial and capitalist practices lack the immediate drama of extreme weather events and thus often escape media attention despite their long-term severity.

Alongside media products that minimize or deny climate change, recent decades have seen a proliferation of apocalyptic media narratives. Notably, climate fiction (cli-fi)[15] films produced in Europe and the United States frequently depict apocalyptic futures—or presents—where the apocalypse is often portrayed as an inescapable endpoint (Meschiari, 2019). These narratives, which respond to what Amitav Ghosh (2016) describes as an 'imaginative and cultural crisis,' displacing ecological catastrophe into dystopian projections. However, they often reproduce normative biases: centering white, heterosexual, cisgender male protagonists, and reinforcing ethnic, racial, and gender stereotypes. Additionally, such films frequently anthropomorphize nature as a vengeful adversary rather than portraying it as a damaged system (Xausa, 2020).

Despite their shortcomings, these cinematic representations shape both individual cognition and collective imaginaries (Milkoreit, 2017). By offering socio-climatic imaginaries—visions of what might be—they help construct a sense of the future—whether desirable or dystopian. Many cli-fi works echo Roy Scranton's (2015) claim in *Learning to Die in the Anthropocene*: that confronting climate change means coming to terms with the inevitability of some form of societal collapse. This sense of doom has been vividly dramatized in numerous films over the past decade, including *Snowpiercer* (2013), *The Day After Tomorrow* (2004), *Don't Look Up* (2021), and *Dune* (2022).

As Eddie Yuen (2012) observes, however, the language of environmental catastrophe in such films often aligns with Hollywood's apocalyptic conventions. This approach strips these narratives of meaningful ethical content and radical critiques of the political, economic, and social structures underpinning the climate crisis (Xausa, 2020). By emphasizing spectacle over systemic critique, these films risk reinforcing a sense of fatalism rather than inspiring actionable change.

Bruno Hodder and Martin (2009, p. 17) expresses concern about the psychological toll of relentless ecological alarm, asking, 'How can we not feel intimately roiled with anxiety at not knowing how to find an answer?'

[15] Climate fiction is a subgenre of science fiction that addresses the issue of climate change through dystopia and apocalyptic and post-apocalyptic science fiction, describing a future in which humans must survive on a planet that has undergone drastic changes. The term is generally attributed to freelance journalist and climate activist DanBloom in 2007. Bloom had used the term to describe his novella Polar City Red, a post-apocalyptic story of climate refugees in Alaska set in 2075, which was not commercially successful.

This pervasive sense of unease, stoked by implausible apocalyptic narratives, often does not spur action—but instead fosters paralysis. The culture industry further amplifies this dynamic by perpetuating what Danowski and Viveiros de Castro (2017) describe as 'perplexing panic,' and 'general amnesia.' These mechanisms align with Klein (2007) concept of the 'shock doctrine' and 'disaster capitalism,' wherein crises are systematically exploited not to address systemic issues but to ram through neoliberal agendas (Fig. 3.1).

Fig. 3.1 Atlas of the New World. What will the atlas on which future generations will study look like? How can a photograph show the future? These questions led to the idea of the Atlas of the New World. Photography is a medium that usually depicts the present. Atlas shows the future in relation to the world we live in today. With a special slide projector, Atlas physically projected onto the landscape the image of its possible transformation. The result is complex and dreamlike photographs that become a metaphorical projection of a not-too-distant future. In the picture: Mozambique is among the most vulnerable African countries to natural disasters. Coastal erosion, frequent droughts, devastating cyclones and floods are episodes that now alternate and recur every year, affecting and destroying entire villages and compromising agricultural harvests. Courtesy of the photographers Giulia Piermartiri and Edoardo Delille

Conflicting Imaginaries of Mobility (in)Justice

As discussed, migration is a structural phenomenon whose definition is shaped by temporal, and socioeconomic contexts and is significantly influenced by media as the infrastructure that materially and symbolically constitutes the world. In contemporary society, both social interaction and cultural reproduction are mediated through various forms of media. Whether analog or digital, media play a crucial role in constructing individuals' perceptions of reality and in creating shared imaginaries and social representations. By influencing what and how we think, traditional and digital media—alongside a diverse array of institutions, actors, sources, tools, and communicative practices that coexist rather than replace one another—contribute to the formation of a collective understanding of the world (Couldry, 2010). These representations may alternately reinforce fear of the 'other' and legitimize their criminalization, or foster curiosity and empathy toward different cultures and places.

Migration has been an integral aspect of human existence, serving as a fundamental mechanism through which humanity adapts to diverse social and physical environments (Bacci, 2015). In recent decades, however—what Castles and Miller (1993) term 'the age of migration'—migration has emerged as a central parameter of global order, profoundly influencing governmental policies and shaping public imaginaries concerning the right to cross borders. Historically characterized by predominantly unidirectional flows from the Global South to the Global North, driven by colonial and postcolonial transformations as well as industrialization, contemporary migration has become increasingly multifaceted. Its causes now span existential, economic, and environmental factors, and its flows encompass both regional and global movements. Despite this diversification, migration continues to reflect the asymmetrical yet interconnected geographies of transnational mobility.

Far from being an issue concerning the US or Europe alone, as Western media often claim, migration is differentially but obstinately global. According to the *World Migration Report 2024* by the International Organization for Migration (IOM), migration patterns continue to reflect a strong regional focus (as for the three previous: 2018, 2020, and 2022 editions). Approximately 31% of migrants remain within Asia, 10% within Africa, 26% within the Americas, and 30% within Europe. Furthermore, the majority of displaced persons—73%—seek refuge in neighboring countries within the Global South, underscoring the regionalized nature of displacement (IOM, 2024). These trends reflect the interconnected yet asymmetrical dynamics of migration and displacement, shaped by socioeconomic, political, and environmental dimensions.

While migration varies in cause and direction, it remains fundamentally shaped by systemic inequalities and transnational mechanisms of control. Australia's offshore detention system, for instance, characterized by indefinite detention of asylum seekers on Nauru and Manus Island, finds parallels in migrant camps on Greek islands and in Albania. These facilities, in turn, mirror the punitive treatment observed at the US southern border, where asylum seekers face similarly inhumane conditions and systemic challenges (Molnar, 2022).

Over the past decade, global preoccupation with immigration has been fueled by economic precarity, armed conflict, and globalization—even in countries where actual migrant arrivals have declined. In Europe, this anxiety peaked in 2015, when over a million people crossed the EU's southern borders. This unprecedented movement was swiftly framed by Western media as a 'crisis,' with headlines and social media amplifying narratives of disruption and emergency (Binotto et al., 2016; Vaughan-Williams, 2015). Such portrayals polarized public perceptions, oscillating between humanitarian empathy and securitarian rhetoric that often dehumanized migrants and refugees.

Since 2015, migration has remained central to the EU policy agenda, driven by perceived threats from 'illegalized' migration and the humanitarian imperative to uphold the rights of people attempting to cross borders (Chouliaraki and Musarò, 2017). Media have played a key role in framing migration and asylum as a 'state of exception' (Agamben, 1998), requiring emergency legislation and intervention. These portrayals often distort empirical realities, inflating a sense of crisis while obscuring the structural conditions behind displacement.

As Calhoun (2010) argues, in the last decades 'the term emergency became a sort of counterpoint to the idea of global order. Things usually worked well, it was implied, but occasionally went wrong. Emergencies were the result, and they posed demands for immediate action' (p. 17). Emergencies are therefore constructed as part of a broad social imaginary in which they are seen as exceptions to normal social life and global organization: sudden, unpredictable, and carrying strong moral imperatives for immediate action. Consequently, what we have nowadays is not one large emergency dismissed as an exception, 'but innumerable smaller ones still treated as exceptions to an imaginary norm but repeated so frequently as to be normalized' (Calhoun, 2010, p. 20). This 'crisis imaginary' has been further intensified by the COVID-19 pandemic, during which migrants were frequently portrayed as disease-carriers. The health emergency legitimized heightened border control within and around Europe, legitimizing further border externalizations (Giacomelli, 2020; Giacomelli et al., 2021; Giacomelli & Walker, 2022).

Institutional and political actors have stoked public fears and security concerns, promoting emergency narratives, militarized border control, and surveillance policies which in turn have generated a fertile breeding ground for xenophobic, populist reactions. Eurocentric media discourses have often portrayed the arrival of refugees and asylum seekers as catastrophic, consolidating the idea of migration as both a humanitarian and security 'crisis.' This framing—with its securitized, depoliticized, and technocratic approach—has helped populist, right-wing political parties amplify their anti-immigration agendas (Wodak, 2015).

Here, Rancière's framework once again helps us understand how toxic narratives are constructed to prioritize certain voices over others: 'Politics revolves around what is seen and what can be said about it, around who has the ability to see and the talent to speak, around the properties of spaces and the possibilities of time' (Rancière, 2009, p.13). Media and narratives play a crucial role in shaping and reinforcing the *distribution of the sensible*—the ways in which what is visible, audible, and thinkable in society is structured. Media, in its various forms, acts as a powerful tool in both maintaining and challenging the established sensory order. By controlling what is represented, who is represented, and how they are represented, the media shapes public perception and influences which narratives are deemed legitimate or worthy of attention. Thus, even if the 'crisis' is symbolically constructed, its consequences are very real (Giacomelli & Walker, 2022). Insofar as they decontextualize the historical phenomenon and dehumanize migrants and asylum seekers, these narratives legitimize unequal power relations and mask unjust applications of the right to move (Goodman et al., 2017). Within this broader context, new manifestations of poverty, social exclusion, spatial segregation, and gender inequalities are increasingly emerging on a global scale.

One of the most significant developments during the peak of this so-called migration 'crisis' has been the emergence of what several scholars call 'humanitarian securitization' (Chandler, 2014; Musarò, 2017). This regime of border governance intertwines national security concerns—such as protecting against perceived threats posed by migrants—with a simultaneous framing of migrants as precarious and vulnerable populations in need of care. As Walters (2011, p. 139) explains, this dual dynamic manifests as a regime of both openness and enclosure, where state borders and entry points are transformed into 'zones of humanitarian government.' Within these zones, humanitarian securitization operates through networks that mediate between risk management and compassionate engagement, casting migrants alternately as 'risks' or even 'terrorists' requiring containment, and as 'victims' in need of care (Chouliaraki & Georgiou, 2022, p. 11).

Dominant Eurocentric discourses about 'humanitarian securitization' are shaped by persistent myths and misconceptions. One is the exaggerated emphasis on South-North migration flows, which is often cast as uncontrollable 'floods' or an all out 'invasion.' In reality, intra-regional migration is far more common than inter-regional migration (Ionesco et al., 2017). Another pervasive stereotype portrays migrants as predominantly single, Black, male, Muslim, unskilled, and dangerous—an image turned into an easy scapegoat for intolerance and hate speech. This stereotype is increasingly challenged by the 'feminization' of international migration. Over the past two decades, the number of migrant women entering the European Union has steadily risen, along with growing recognition of their economic contributions to European societies.[16] Scholars such as Anthias and Lazaridis (2020), Pinelli (2019), and Decimo (2024) underscore how women—long depicted solely as vulnerable subjects or economic burdens—are now acknowledged as active contributors to Europe's economic and social fabric through their assets, networks, and agency.

Despite comprising only a small portion of total EU mobility, irregular border crossings receive disproportionate media coverage. This over-mediatization creates a significant gap between public perception and the reality of migratory dynamics (ISMU, 2021). The resulting misrepresentation has fueled populist agendas, elevating migration to a politicized flashpoint. By framing migration as an existential crisis requiring urgent intervention, media narratives often amplify securitization rhetoric and

[16] See the data available on the website of the Migration Data Portal (https://www.migrationdataportal.org/about) managed and developed by IOM's Global Migration Data Analysis Centre (GMDAC).

fear at the expense of more nuanced discussions about systemic inequalities and long-term solutions.

Building on the concept of mediatization as a process through which 'the media exert a particularly dominant influence on other institutions' (Hjarvard, 2008), we come to see how the symbolic dimension of the border is shaped by the binary logic of 'victim' versus 'terrorist.' This dichotomy underscores how media representations construct and reinforce polarized narratives that influence public perception and institutional practices surrounding migration and border governance.

The black-and-white portrayal of migrant identities as either 'desirable' or 'undesirable' is linked to what Chouliaraki and Musarò term the 'mediatized border' (Chouliaraki and Musarò, 2017): a techno-affective network of migrant and refugee mediations, where emotions of fear and empathy coexist through digital connectivities and ritualize relationships with the other through discourses of difference and superiority. Despite their seemingly different mandates, the military and humanitarian structures coexist, the former charged with controlling and excluding, the latter with providing migrants with medical aid, shelter, nutrition, and information.

Whether territorial or representational, the border, as Vaughan-Williams puts it, is always 'a process of bordering' that 'seeks to rhetorically identify and control the (very) mobility of certain people, services, and goods that operate around its jurisdiction' (Vaughan-Williams, 2015, p. 6). As a technologically driven process of 'rhetorical identification and control,' bordering thus systematically produces its own 'discursive or emotional landscapes of social power' (Paasi, 1996, p. 63).

The victim/terrorist dichotomy is a core discursive feature of Europe's border governance. As it crystallizes a logic of humanitarian securitization, it merges affective appeals to compassion with militarized security imperatives. This symbolic bordering process was starkly illustrated by Italy's *Operation Mare Nostrum* (2013–2014), which revealed how militarized humanitarianism functions as an apparatus of selective inclusion—rescuing while simultaneously policing mobile populations. The operation's affective politics relied on a gendered narrative of heroic military intervention, where naval forces were framed as both empathetic saviors and sovereign gatekeepers. Public discourses emphasized the gratitude of rescued migrants ('victims') toward European actors while portraying prospective border-crossers as latent threats, thereby naturalizing the hegemonic security logic over humanitarian principles.

This regime of militarized care produces a paradoxical border epistemology: a zone of conditional permeability where rescue operations coexist with exclusionary practices. As demonstrated in Mare Nostrum's fusion of naval patrols and search-and-rescue missions, the Mediterranean became a theater for performative compassion that ultimately reinforced border externalization (Pezzani & Heller, 2019). The operation's media representations elevated the 'humanitarian-soldier' archetype over substantive engagement with migrants' agency, reducing rescued individuals to passive recipients of European benevolence. Such narratives align with what critical scholars identify as humanitarian governmentality—a biopolitical regime that administers life through hierarchies of grievability (Butler, 2010a), whereby certain populations are rendered deserving of protection while others remain expendable.

Migrant subjectivity under such paradigms oscillates between hypervisibility (as humanitarian objects) and erasure (as political subjects). As Chouliaraki and Musarò (2017) contend, the border's 'human' remains an ambivalent figure—simultaneously invoked through sentimental appeals to shared vulnerability and negated through dehumanizing security protocols. This tension animates EU policies that frame migration management through a risk calculus, privileging maritime surveillance technologies and interception partnerships over rights-based protection frameworks. The result is a humanitarian border that operates not as contradiction but as synthesis: a space where care and control coalesce to reproduce exclusionary hierarchies under the guise of ethical responsibility.

The Securitarian Frame: Migrants as Threats

The characterization and treatment of migrants varies depending on historical periods, ruling governments, and—most significantly—media portrayals. This process is inherently ambivalent, as migrants are simultaneously framed as both '*risky subjects*' and '*subjects at risk*' (Aradau, 2004; Garelli & Tazzioli, 2017). As '*risky subjects*' they are perceived as unwelcome threats to European societies and values. Conversely, as '*subjects at risk*,' they are depicted as victims—either of their countries of origin or of Europe's stringent border control policies, commonly referred to as 'Fortress Europe' (Chouliaraki & Georgiou, 2022).

In the first narrative, the portrayal of migrants as security threats legitimizes extreme measures and policies that were once deemed unjustified or inhumane. This approach, often described as 'crimmigration,' not only symbolically frames migrants as societal threats but also facilitates legal and extralegal efforts to exclude them from European territories (Gatta et al., 2021). Consequently, public fears are addressed through the construction of both physical and symbolic border fortifications, often at the expense of human rights. A silent manifestation of this approach is the outsourcing of migration management to non-EU countries via agreements such as the EU-Turkey deal (2016), Italy-Libya Memorandum of Understanding (2017), and similar arrangements with Tunisia, Senegal, and Albania. Although intended to deter migration, these policies have resulted in significant policy failures—including the refusal of several European nations to ratify the *Global Compact on Migration*—exposing systemic distortions and shortcomings within current migration governance.

Fear of migrants is not unique to modernity. Populist media commentators and politicians often conflate the identity of migrants with external threats, portraying migration as a temporary disruption to a fixed national identity. However, such representations contradict historical realities. Anatomically modern humans have spent 98% of their existence as nomads, meaning that mobility in search of a better life is fundamental to the human experience. Nonetheless, the mobility of certain populations is cast as deviant in relation to a dominant residentialist norm, subjecting them to scrutiny, suspicion, and surveillance.

As noted in the previous chapter, aspirations to move freely are not equally granted. Rather, differential mobility regimes impose restrictions and criminalize movement. Papastergiadis (2012) critiques the residentialist mode of culture, arguing that its negative stance toward migration reflects a broader ambivalence toward mobility. He identifies 'kinetophobia, or fear of movement, as manifesting in various forms, most notably in racist scapegoating that attributes social upheaval to vulnerable agents of movement' (p. 39). This residentialist view has produced adversarial positions against hybridity and cultural transformation—privileging identity over diversity, stasis over movement, and one-sided surveillance over mutual recognition. Migrants, it follows, are conceptualized as subordinate political figures within the static foundations of state sovereignty (Nail, 2016).

Across the Global North and beyond, a generalized 'atmosphere of hostility' has emerged, characterized by the erosion of social protections and increased precariousness for those marked as outsiders and denied access to work, housing, services, and education. This hostility is exacerbated by the global rise of nationalist and populist political forces that scapegoat migrants—defined not only by foreign citizenship but also by racialized otherness–as prime targets of public anxiety. Through transnational cooperation in surveillance, risk analysis, and deportations, EU and non-EU border policing has become an expansive regime that operates before, at, and beyond the border, transforming entire transit environments into spaces of hostility and mobility deterrence. This and similar strategies of border control have linked the 'natural' borderlands located at the fringes of the Global North to urban geographies in Europe, the United States, Australia, and beyond. As Walters et al. (2022) note, what emerges is not a singular hostile environment, but a vast multiplicity of such environments that span 'natural' and civic spaces alike.

The 'securitization of migration' has shaped our common understanding of what or who represents a (potential) threat. As Bauman argues, 'crisis has become a kind of politically correct code name for this phase of the eternal struggle waged by opinion-makers to conquer and subjugate minds and hearts' (Bauman, 2016, p. 3). This is echoed in the 52nd report by the CENSIS[17] Institute for Social and Economic Research (2018), which diagnoses Italy as suffering from both fear of the future and fear of migrants, introducing the term 'psychic sovereignty,' to describe a generalized and painful perception of losing national control accompanied by an exacerbated fear of the 'other'—primarily immigrant people. The following year's report (2019)[18] found that nearly 7 out of 10 Italians believe that hatred, intolerance, and racism toward minorities in the country have increased over the previous year.

The distorted representations of migration have fueled a 'moral panic' (Cohen, 2011) resulting in waves of hate crimes and discriminatory speech (Faloppa, 2020). Harmful and dehumanizing narratives on migration have permeated political discourse and mainstream media across numerous countries. Such narratives have scapegoated migrants for deep-rooted societal problems and fears, often for electoral or financial gain.

In their study of metaphors employed by European and US media during the so-called 'refugee crisis' (2015–2016), Arcimaviciene and

[17] http://www.censis.it/rapporto-annuale/52%C2%B0-rapporto-sulla-situazione-socialedel-paese2018-0.

[18] http://www.censis.it/rapporto-annuale/53%C2%B0-rapporto-sulla-situazione-socialedel-paese2019.

Baglama (2018) identified a pattern of metaphorical expressions in journalistic discourse deemed to be highly stigmatizing and dehumanizing. These expressions often promote the 'objectification' or 'animalization' of migrants, utilizing terms such as 'sheltered,' 'trapped,' and 'beached.' Such language fosters an image of migrants as instinctual and irrational beings, effectively portraying them as entities detached from the human community. Furthermore, the metaphorical framing of migrants as natural disasters—such as 'flows,' 'spills,' and 'waves'—amplifies perceptions of mass movement as chaotic and threatening. Although these metaphors predate discussions of climate-induced migration, their continued use has contributed to accentuating the unpredictable and forceful nature of migratory movements, potentially instigating feelings of fear and insecurity in populations while exacerbating animosity toward migrants and refugees.

Humanitarian Frame: Migrants as Victims

Despite its internal diversity, the literature on the migration 'crisis' consistently highlights how symbolic borders materialize through harmful narratives that misrepresent migrants—frequently casting them either as victims or villains (Crawley et al., 2016; Zaborowski & Georgiou, 2019). As Malkki (1996) observed nearly three decades ago, migrants are often caught within oppositional webs—cast either as helpless sufferers or dangerous threats—consequently denied agency, voice, and dignity. Reviving classic orientalist paradigms (Said, 1979), these portrayals silence and objectify migrants while stripping them of their individuality and biographical depth. Thus primitivized, they are reduced to what Agamben (1998) calls 'bare life'—a condition in which their existence is acknowledged only at the level of biological survival, devoid of political or human recognition. Within this frame, migrants become ambivalent figures whose lives may warrant protection but whose deaths often remain unprosecuted and unmourned. Agamben's concept of 'bare life' illuminates the exclusionary logic of sovereign power, where migrants are relegated to a zone of exception—present within legal and political systems, yet denied all protections. Such narratives reinforce systemic inequalities and perpetuate a framework that dehumanizes migrants and legitimizes their marginalization within public discourse and institutional practices.

Building on Judith Butler's theory of differential public grieving—'an ungrievable life is one that cannot be mourned because it has never lived, that is, it has never counted as a life at all' (Butler, 2010b, p. 9)—Martina Tazzioli (2020) develops the concept of 'geographies of (un)grievability.' Analyzing the positioning of specific populations and regions within global systems of governance and labor—often characterized by restricted access to rights and protections—Tazzoli reveals how media representations contribute to constructing certain lives as invisible and undeserving. These portrayals frequently dehumanize migrants, refugees, and marginalized people, relying on tropes of danger, desperation, and helplessness. Such framings significantly shape public opinion and policy, reproducing geographies of exclusion: regions like border zones, refugee camps, or impoverished areas are depicted as spaces of suffering, while the individuals within them are denied visibility in terms of rights, justice, or solidarity.

For example, media coverage frequently sensationalizes migration crises without addressing their structural causes—such as imperialism, conflict, or economic inequality—thereby diverting attention from systemic injustices that perpetuate suffering. In doing so, a reductive narrative is reinforced that emphasizes immediate crises while obscuring broader historical and political contexts. These representations marginalize vulnerable populations and undermine meaningful engagement with the root causes of displacement and inequality.

Shedding light on the intimate link between cosmopolitan solidarity and the humanitarian imaginary in our contemporary society, Lilie Chouliaraki identifies as 'theatrical' that 'configuration of practices which use the communicative structure of the theater in order to perform collective imaginations of vulnerable others in the West, with a view to cultivating a longer-term disposition to thinking, feeling and acting toward these others' (Chouliaraki, 2013, p. 45). Such practices of performance, which include the early genres of posters, paintings, novels, appeals, celebrities, concerts, and the news, make up a theatrical structure that mundanely produces imaginations of solidarity—in a sense the field of action Arendt (1998) called 'our common shared world.' This definition also lies at the heart of Taylor's conception of *Modern Social Imaginaries* (Taylor, 2003) as spaces of communication consisting precisely of 'those images, stories and legends' through which 'people imagine their social existence, how they fit together with others, [...] and the deeper normative notions and images that underlie these expectations' (p. 106).

The imperative to alleviate the suffering of distant *others* is often framed as a humanitarian duty, calling for moral action irrespective of geographical or cultural proximity. Humanitarian solidarity transcends national borders, emphasizing the ethical obligation to recognize the humanity of those affected by crises—whether due to war, poverty, or displacement. It stresses the need for compassion and action across borders, even when the affected populations are far removed from our most immediate context.

However, this mode of solidarity can be vexed by political, economic, and media-driven narratives that shape how and when vulnerable others are seen as deserving of help, often privileging certain groups over others. On the one hand, as Chouliaraki (2017) argues, 'the most meaningful vision of cosmopolitanism relies on communicative acts that recognize the humanity of vulnerable others and engage with their vulnerability without demanding reciprocation' (p. 53). From here it follows that the stories of suffering become primers for the exercise of our 'citizenship of the world,' for the sentimental cultivation of 'our moral community,' and for the training of our sympathetic, moral imaginations (Rorty, 1989, p. 75). On the other hand, when distant vulnerable others try to cross the border and get closer, they automatically become unwanted, uninvited, potential criminals. This paradox reveals how humanitarianism is inscribed in systematic patterns of global inequality and in hierarchies of place and human life (Fassin, 2011). Here we can see again the tensions among different scales of mobility (in)justice. These hierarchies divide the world into zones of Western comfort and safety and zones of non-Western need and vulnerability showing that the politics of compassion is a politics of inequality (Mezzadra & Neilson, 2013).

As previously hinted, there are often contradictions and conflicts between the different scales of justice. Indeed, humanitarianism can be viewed as a historically specific articulation of cosmopolitan solidarity, 'which acts directly on the Global South through specialized institutions, yet seeks legitimacy in the West through a communicative structure that disseminates moral discourses of care and responsibility' (Chouliaraki, 2013, p. 27). Nevertheless, if we explore the multifaceted interaction between social inequality and humanitarianism in a globalized world within the framework of symbolic bordering, it becomes evident that the juxtaposition of humanitarian and securitarian narratives construct a 'moral geography of the world' (Musarò & Parmiggiani, 2022) suitable for governing the 'borderlands' (Duffield, 2001) while unifying Europe against the 'Other.' Within this narrative, the 'European' subject values and protects the 'African' victim by positioning him/herself simultaneously as a donor and a controlling agent (Cappi & Musarò, 2022).

The social production of the African and the European continents as binary opposites is the product of a symbolic bordering process (Guiraudon & Joppke 2003; de Haas, 2007). The cultural construction is that of the North which is wealthy, ordered, compassionate, and a natural bestower of a civilizing mission to the underdeveloped, dangerous, and voiceless South. This contrast is embedded in mediated representations of the 'other' and the 'we' based on asymmetry of power (in terms of both agency and dignity). Migrants are often portrayed as an undifferentiated mass unworthy of humanitarian attention; they are removed from an ethical framework and de-individualized, reduced to a simple object of our pity (Boltanski, 1999). Through such representational logics, humanitarian concern becomes a stage for affirming European virtue. Heroic in their aid to the poor, needy, and passive 'others,' European identity is paradoxically reinforced along the axis of humanitarianism and control.

Migrants are thus systemically presented as objects of charity. Treated as having no social or personal history of their own, they become 'non-persons' (Dal Lago, 2012). In addition, 'illegal' migrants, asylum seekers, and refugees—subsumed within a single story of illegitimate(d) mobility—are represented as outside the common history that binds countries of origin, transit, and arrival. In fact, no explanation is given about the multiple factors that brought them to where they are, the political, environmental, economic, or social context that drive them, or the different ways in which they participate in arrival countries in roles other than that of a migrant (Cappi et al., 2021). Not only does this representation of migration as a series of emergencies inhibit understanding of the causes of the disparities between the 'spectator' and the 'victim'—it also increases the distance between 'us' and 'them' (Musarò & Parmiggiani, 2014).

Entrepreneurial Integration Frame

In line with the processual conception of the (digital) border—which Chouliaraki and Georgiou define as a 'versatile regime of power whose technologies and discourses of control follow migrants as they move and settle inside the nation that receives them' (Chouliaraki & Georgiou,

2022, p. 58)—we can focus on the process of bordering as a process of systematic exclusion of racialized bodies not only at crossing points but also inside Western nation-states (Yuval-Davis et al., 2018). This helps us to conceptualize the inner border, where migrants remain subjected to data surveillance and physical control (akin to that exercised at the outer frontier), as well as to media narratives that cast them as part of an ongoing 'crisis.' In such accounts, migrants are still perceived as problematic (whether as a security threat or burdens on welfare systems), but are simultaneously cast as essential contributors to the economic vitality and growth of host societies. They are thus portrayed not only as a challenge but also as a resource that enhances the productivity and resilience of the host country.

Particularly in the cities, where migrants often settle, the frame of humanitarian securitization is supplanted by a more strategic regime of power, that of 'entrepreneurial securitization,' which Chouliaraki and Georgiou define as a 'regime that situates the surveillance of migrant lives no longer within technologies of humanitarian care but within the market rationalities and narratives of entrepreneurial competition' (Chouliaraki & Georgiou, 2022, p. 59).

Within this frame, host societies increasingly valorize migrants as entrepreneurial subjects—drivers of innovation, job creation, and economic growth. Migrants are recognized for their resilience and resourcefulness, often starting businesses in sectors that local populations may overlook or struggle to breach. While commonly exploited in low-wage sectors such as agriculture, construction, domestic work, care, and cleaning (Omizzolo, 2019), their entrepreneurial spirit is particularly valued in areas such as retail, technology, hospitality, and food services, where they introduce fresh ideas and market strategies. Particularly in Western nations with aging populations or struggling economies, migrant entrepreneurs are framed as essential to revitalizing local economies. According to the 2022 IOM report, 74% of all international migrants are of working age (20–64 years) and they exhibit higher entrepreneurial activity compared to native populations—likely a consequence of having to creatively respond to exclusion from formal labor markets combined with new economic opportunities for trade across diasporic and transnational networks (Yamamura & Lassalle, 2020).

Although migrant businesses often face discrimination based on their nationality or legal status, migrant entrepreneurship is increasingly celebrated as a means of promoting social integration (Baycan-Levent &

Nijkamp, 2009), transforming derelict neighborhoods or streets into vibrant local markets (Rath & Schutjens, 2019), and enhancing the global competitiveness of host economies (van Delft et al., 2000). This recognition challenges the notion of migration as a societal burden and instead highlights migrants' agency in reshaping and strengthening the economies and social fabric of their new homes.

However, seen within the dominant discourse of neoliberal citizenship, migrant entrepreneurial success is not driven by economic benefit alone; rather, it is a more complex matter informed by ontological, socio-cultural, and legal implications. The status of many asylum seekers lies between full precarity and the promise of recognition. They experience extended periods of legal limbo in which they can only access some tenuous forms of informal work (while being asked formally to not work at all) until they receive (if indeed they ever do) a temporary residence permit. For such individuals, employment is more than an income: it represents a step toward safety. It means being able to claim recognition as respected residents of western cities, and thus move closer to the possibility of eventually achieving the legal status of citizen.

This underscores the importance of what Jessop and Sum (2000) call 'narratives of enterprise' within a neoliberal logic that elevates the norms of individualism and self-autonomy into prerequisites for granting migrants legal status and treating them as legitimate members of their new communities. This means that these narratives tend to ignore the structural conditions of unemployment and existential precarity that migrants face upon arrival.

The contradictions of this entrepreneurial frame are evident in the host country's categorization of migrants as either 'deserving' or 'undeserving.' While entrepreneurial migrants are celebrated as beneficial to the economy, others are marginalized and stigmatized. We can observe the tensions between positive and negative frames of migration in hundreds of integration programs set up to help skilled immigrants become self-sufficient. But what kind of integration is being promoted here?

Although integration spans all facets of migrant life—from education and employment to cultural belonging—Laura Zanfrini (2016, p. 10) points out how the typologies used by destination countries to organize migration flows and categorize them administratively, reflect the interests of host society far more than migrants' lived realities.

To achieve inclusion within host societies, it is essential to address the concept of 'internal acceptance,' which is deeply shaped by political and social dynamics that vary across historical periods and are influenced by factors such as colonialism, racism, and state power. Ambrosini identifies two key dimensions: recognition and acceptance. Both are closely tied to the notion of social visibility, which has been significantly impacted by the mediatization of migration in recent years (Musarò & Parmiggiani, 2022). Ambrosini intersects the dimensions of acceptance and recognition to outline four ideal-typical scenarios in native-foreigner relations: *exclusion, stigmatization, tolerance, and integration* (Ambrosini, 2014, pp. 39–43). *Exclusion* arises when migrants lack both legal authorization and social recognition, resulting in overt hostility and the label of 'illegal immigrants.' *Stigmatization* occurs when migrants possess legal authorization but face significant social rejection from locals. This is particularly evident in the case of 'socially undesirable' minorities such as Roma, Sinti, or refugees, who are frequently perceived as foreign and threatening despite holding regular residence permits.

Conversely, *tolerance* emerges when migrants lack formal authorization but receive a degree of social recognition. Ambrosini here gives the example of undocumented immigrant women working invisibly as caregivers or maids. These subjects fit into the category of undocumented migrants deemed 'deserving' (Chauvin & Garcés-Mascareñas, 2014), whose presence is legitimized through their labor—typically underpaid, precarious, socially penalized, and carried out with minimal human rights protections (Ambrosini, 2020). This dynamic reflects the paradox that migrants are 'in demand but not welcome' (Zolberg, 1997, p. 143), exposing the tension between economic necessity and societal rejection (Bolaffi, 2001, p. 61).

Integration, finally, occurs when formal authorization aligns with social recognition. This concept remains theoretically ambiguous and is often employed with varying or conflicting meanings. The European Agenda for the Integration of Third-Country Nationals (2011) defines integration as a 'dynamic, two-way process of mutual adjustment between immigrants and host societies' (p. 2). In academic discourse, integration functions as a conceptual bridge between migration flow studies and public policies addressing migration (Boccagni & Pollini, 2012). Sociological theories on integration range from assimilationism (Park & Burgess, 2014), multiculturalism (Colombo, 2011), interculturalism (Benhabib, 2002), functionalism (Bramanti, 2011), and transnational perspectives that frame migration as 'globalization from below' (Ambrosini, 2009). Absent a universal definition, integration must be viewed not as a progressive line, but as a multifaceted process unfolding unevenly across different dimensions and temporalities.

In conclusion, dominant integration paradigms often overlook what Sandel (2020) refers to as 'the dignity of work'—a concept encompassing the full spectrum of human potential. As a result, integration tends to privilege those migrants who succeed as entrepreneurial subjects, while further marginalizing those unable to meet these criteria (Fig. 3.2).

Fig. 3.2 A People's Archive of Sinking and Melting is a collection of materials contributed by people living in places that may disappear because of climate change. From upper left: Film reel, Audio tape paper insert, Drawing on graph paper, FRYSLAN postcard, One dollar bill (USD), Audio tape case, Dirty paper (2), Kodachrome film box, Queen, We Will Rock You, Wall plaster chip, slide. Caption: All materials in the New York collection were collected by Adriane Colburn and Tyler Henry during the cleanup of Smack Mellon from their basement artist studios after Superstorm Sandy. November 5, 2012. Image: Balkin, et al., A People's Archive of Sinking and Melting, 2012-ongoing New York Collection contributed by Adriane Colburn and Tyler Henry with Golnar Adili. Image Credit: Mary Lou Saxon Courtesy of the Archive

Conclusions

As Barca writes (Barca, 2020), narratives do not kill per se, but they can invisibilize both perpetrators and victims. It is therefore unthinkable to address climate and mobility justice without critically engaging with the ways in which dominant narratives deny colonial, gender, class, and species relations.

In this chapter, drawing on Rancière's notion of the distribution of the sensible to capture the relationship between narratives, perception, and politics, we have examined the dissonance between expert knowledge and populist rhetoric in shaping public debate. Through an analysis of media framing practices, we have shown how certain discursive techniques are used to deny scientific facts, legitimize policy responses, and obscure alternative viewpoints—ultimately reinforcing existing power structures and inhibiting challenges to the status quo.

The gap between scientific discourse on climate change and populist interpretations, as well as the gap between statistical data on migration and its politicized representations in media, directly informs citizens' attitudes and imaginaries. Public discourse as presently constructed and amplified by dominant forms of media, conditions how global warming and migration are understood, translating alarmism into an apparent objective reality.

In exploring the links between climate change, mobility regimes, and (in)justice (Latour, 2020), we have also highlighted the traps—epistemic, communicative, and emotional—that prevent public consensus on a common definition of justice and thus paralyze action. Studies of media coverage have long shown that the media play a central role in shaping climate policy and public opinion. Spectacularized narratives of climate change amplify emotional responses at the expense of sustained engagement (Bonati, 2015). Understanding how narrative frames produce 'climate imaginaries' helps explain not only what we know about the crisis but also how we feel and act—or fail to act—in response. As Greta Thunberg puts it, it also determines what we don't do about it.

While scientists continue to confirm IPCC's worst-case predictions (with deadly wildfires, horrific flooding, and record-high temperatures daily and globally at stake), US President Donald Trump continues to downplay the threat of climate change. His latest attack includes withdrawing support from any research that even mentions the word 'climate'.[19] He has also launched a purge on government websites hosting climate data, in an apparent attempt to make the evidence disappear. Senior officials in agencies like the Environmental Protection Agency (EPA) were instructed to avoid using certain words in public documents, press releases, and official communications, in a bid to shift the narrative and avoid legitimizing the scientific consensus on climate change. At the same time, despite widespread skepticism, Elon Musk continues to promote the idea that it will be possible and essential for humans to colonize our neighboring planets as a form of insurance against devastation on Earth.

These positions work to bolster anti-science extremists—including climate science deniers—and reinforce the cognitive and the communicative traps so often shaped by the techno-utopian and dystopian-apocalyptic scripts we outlined prior. They also legitimize the mass climate deception campaigns masterminded by fossil fuel and oil corporations, while reversing environmental regulations and preventing new policies from being enacted.

Looking beyond Trump's antagonism toward environmental and climate policy, we also critiqued the *Loss and Damage* mechanism for its failure to address the root causes of climate change. As several theorists and activists point out, the inability to recognize the material and social limits of growth is one of the main false premises (and promises) underlying most negotiations on the fairness of international treaties on climate change (Derman, 2020; Vanhala, 2016). These negotiations typically adopt the distributive logic of the equal shares approach—also termed equal atmospheric shares. It consists of dividing a shared resource into basic units and ensuring that everyone has their fair share. From this perspective, justice is about specifying 'who should take responsibility,' and 'who should perform and how much' (Caney, 2016, p. 24).

Framing the atmosphere as the basis for a reckoning of rights and obligations not only contributes to its commodification but also solidifies a zero-sum logic, where one party's gain inevitably translates to another's loss. This approach treats the atmosphere as a finite resource that must be divided up and neglects the broader, shared responsibility to protect it for future generations, as well as for plants, rivers, oceans, and animals.

[19] https://www.nytimes.com/2025/07/31/climate/trump-climate-skeptics-science-report.html.

Further, by focusing on quantifying and exchanging atmospheric rights, it limits the cosmopolitan political arena to the confines of the human polis. In so doing, it reproduces the conflicts between the unequal international scales of justice, reinforcing the idea that environmental benefits can only be secured through competition rather than collective stewardship, and stymying the potential for fairer collaborative solutions.

To challenge the framing of 'crisis' in Western public discourse, we have argued that the diffusion of apocalyptic and dystopian narratives—framed as 'emergency'—constructs an ecological imaginary shaped by urgency but devoid of shared meaning. As Van Aken (2020, p. 76) states: 'We can't find [or won't find] shared words and frameworks, so the climate crisis remains in a shouted public silence from emergency to emergence.'

Human mobility is similarly caught in this 'crisis imaginary.' As demonstrated in the second part of this chapter, the representation strategies and discursive practices enacted by a wide range of State and non-State actors have been presenting migrants, and particularly the border-crossing of irregular migrants, as an 'emergency' to be managed in terms of a wider social, cultural and political 'crisis' (Bigo, 2002; Mezzadra & Neilson, 2013). The 'migration crisis,' instrumentally constructed by politicians and media actors who benefit from it, feeds on the fears of citizens and deepens divisions by providing static, simplistic answers promoting exclusion. Hence, migration becomes a 'weapon of mass distraction' for political leaders deployed to sideline other prescient issues and protect particular interests.

The very act of naming such phenomena a 'crisis'—especially as a 'European migration crisis'—is itself a discursive maneuver that naturalizes boundary-making. Stripped of historical context such as the backlash of conflicts in Syria, Afghanistan, and the Middle East, as well as the broader trends of migration and the rising numbers of forcibly displaced people worldwide, public discourses in the Global North present migration as a spectacle—an emergency to be managed rather than a consequence of global injustice. Consequently, public anxiety over migration and asylum-seeking is increasingly shaped by the political rhetoric that depicts Europe as an endangered continent besieged by people fleeing conflict or seeking opportunities for a better life.

As we discussed, since the rise of modern nation-states, borders have played an important double role in the ordering society. First, they define territories and delimit the geographies of nations. Second, as social imaginaries, borders shape national identities and world perceptions. From this perspective, processes of bordering are not only social phenomena but also political insofar as they become a strategy to control both real and symbolic space.

While serving as territorial markers, borders also act as mediators of socially constructed processes (Vaughan-Williams, 2015). Borders are practices that produce norms of exclusion across bodies and simultaneously give voice to would-be migrants sustaining projects of geopolitical sovereignty. This process of bordering contributes to normalizing the extraterritorial border by negatively stereotyping migration, or even by criminalizing it. Hence, European, American, and Australian ways of tackling the 'migration crisis,' particularly ones articulated in terms of risk, death, law-breaking, and post-arrival failure—do more than criminalize migration, they partake in the mediatic shaping of a border 'landscape' (Cuttitta, 2015).

Exploring the border as a dialectical site of control and care, we witnessed a process that both defends territorial outputs and protects those it treats as a risk—giving way to the ambiguous regime of 'humanitarian securitization' (Watson, 2011). This regime is a negative consequence of the 'kinetophobic prism' (Papastergiadis, 2012) through which the Global North interprets human mobility from the Global South. This viewpoint is reflected in the discourses propagated by the media, politicians, and border control agencies, which consistently portray individuals fleeing poverty or conflict—seeking a new life in the Global North—within a binary framework of victim versus terrorist. This divisive logic has long been mobilized to vet newcomers between those deserving care and those deemed fit for containment or expulsion.

The kinetophobic prism is also at play in the clash between a vision of cosmopolitanism as solidarity toward distant and vulnerable others—typically found animating the humanitarian field as a whole (Fassin, 2011)—and the particular discourses associated with the operations of humanitarian aid in the context of border control.

This ambivalent and contradictory form of global migration control takes place within the hegemony of a Western epistemology of the earth, which, as Mbembe (2020) argues, views the planet as a site of domination over land and sea circumscribed by borders between nations and peoples. Narratives of migration participate in this logic. Whether in old or new media, they serve the border's function by legitimizing certain discourses of migration over others and engendering the biopolitical power relations of human mobility.

References

Abbas, M. (2012). Climate change as a global political issue. *Atoms for Peace: An International Journal, 3*(3), 219. https://doi.org/10.1504/AFP.2012.046738

Agamben, G. (1998). *Sovereign power and bare life*. Stanford University Press.

Ambrosini, M. (2009). *Un'altra globalizzazione La sfida delle migrazioni transnazionali*. Società editrice il Mulino.

Ambrosini, M. (2014). *Non passa lo straniero? Le politiche migratorie tra sovranità nazionale e diritti umani (1a edizione)*. Cittadella editrice.

Ambrosini, M. (2020). *L'invasione immaginaria: L'immigrazione oltre i luoghi comuni (Prima edizione)*. GLF editori Laterza.

Anthias, F., & Lazaridis, G. (Eds.). (2020). *Gender and migration in southern Europe: Women on the move* (1st ed.). Routledge. https://doi.org/10.4324/9781003085447

Aradau, C. (2004). Security and the democratic scene: Desecuritization and emancipation. *Journal of International Relations and Development, 7*(4), 388–413. https://doi.org/10.1057/palgrave.jird.1800030

Arcimaviciene, L., & Baglama, S. H. (2018). Migration, Metaphor and Myth in Media Representations: The Ideological Dichotomy of "Them" and "Us". *SAGE Open, 8*(2), 2158244018768657. https://doi.org/10.1177/2158244018768657

Arendt, H. (1998). *The human condition* (2nd ed.). The University of Chicago Press.

Armiero, M. (2021). *L'era degli scarti: Cronache dal Wasteocene, la discarica globale*. Einaudi.

Armiero, M., & De Angelis, M. (2017). Anthropocene: Victims, narrators, and revolutionaries. *South Atlantic Quarterly, 116*(2), 345–362. https://doi.org/10.1215/00382876-3829445

Bacci, M. (2015). *Il pianeta stretto*. Il mulino.

Baldwin, A. (2017). Rearranging desire. On whiteness and heteronormativity. In G. Bettini & A. Baldwin (Eds.), *Life adrift: Climate change, migration, critique* (pp. 211–226). Rowman & Littlefield International.

Barca, S. (2020). *Forces of reproduction: Notes for a counter-hegemonic anthropocene*. Cambridge University Press.

Bauman, Z. (2016). *Strangers at our door*. Polity press.

Baycan-Levent, T., & Nijkamp, P. (2009). Planning and Management of Urban Green Spaces in Europe: Comparative analysis. *Journal of Urban Planning*

and Development, 135(1), 1–12. https://doi.org/10.1061/(ASCE)0733-9488(2009)135:1(1)
Benhabib, S. (2002). *The claims of culture: Equality and diversity in the global era*. Princeton University Press.
Berger, P. L., & Luckmann, T. (1990). *The social construction of reality: A treatise in the sociology of knowledge*. Anchor Books.
Beyers, L. (2006). Review: Jane Elliot (2005). Using narrative in social research. Qualitative and quantitative approaches [19 paragraphs]. *Forum Qualitative Sozialforschung/Forum: Qualitative Social Research, 7*(2) http://nbn-resolving.de/urn:nbn:de:0114-fqs0602192
Bigo, D. (2002). Security and immigration: Toward a critique of the governmentality of unease. *Alternatives: Global, Local, Political, 27*(1_suppl), 63–92. https://doi.org/10.1177/03043754020270S105
Binotto, M., Bruno, M., & Lai, V. (Eds.). (2016). *Tracciare confini: L'immigrazione nei media italiani*. FrancoAngeli.
Boccagni, P., & Pollini, G. (2012). *L'integrazione nello studio delle migrazioni: Teorie, indicatori, ricerche*. FrancoAngeli.
Boccia Artieri, G. B., Colombo, F., & Gili, G. (2022). *Comunicare: Persone, relazioni, media (Prima edizione)*. GLF editori Laterza.
Bolaffi, G. (2001). *I confini del patto: Il governo dell'immigrazione in Italia*. Einaudi.
Boland, R. J., & Tenkasi, R. V. (1995). Perspective making and perspective taking in communities of knowing. *Organization Science, 6*(4), 350–372. https://doi.org/10.1287/orsc.6.4.350
Boltanski, L. (1999). *Distant suffering: Morality, media, and politics*. Cambridge University Press.
Bonati, S. (2015). Multiscalar narratives of a disaster: From media amplification to Western participation in Asian tsunamis. *Culture Unbound, 7*(3), 496–511. https://doi.org/10.3384/cu.2000.1525.1572496
Bourdieu, P. (1980). *Le sens pratique*. Éditions de Minuit.
Bramanti, D. (2011). *Generare luoghi di integrazione: Mondelli di buone pratiche in Italia e all'estero*. FrancoAngeli.
Buell, L. (1996). *The environmental imagination: Thoreau, nature writing, and the formation of American culture (2 print)*. Belknap Press of Harvard Univ. Press.
Butler, J. (2010a). *Frames of war: When is life grievable? (1. publ. by verso)*. Verso.
Butler, J. (2010b). Performative agency. *Journal of Cultural Economy, 3*(2), 147–161. https://doi.org/10.1080/17530350.2010.494117
Calhoun, C. (2010). The idea of emergency: Humanitarian action and global (dis)order. In D. Fassin & M. Pandolfi (Eds.), *Contemporary states of emergency: The politics of military and humanitarian interventions* (pp. 18–39). Zone Books.

Canavan, G. (2021). Science fiction and utopia in the Anthropocene. *American Literature, 93*(2), 255–282. https://doi.org/10.1215/00029831-9003582

Caney, S. (2016). The struggle for climate justice in a non-ideal world. *Midwest Studies In Philosophy, 40*(1), 9–26. https://doi.org/10.1111/misp.12044

Cappi, V. (2023). *Immaginare l'altrove nell'epoca dell'Antropocene. Media, confini e cambiamenti climatici*. FrancoAngeli.

Cappi, V., & Musarò, P. (2022). Awareness campaigns to deter migrants: A neoliberal industry for symbolic bordering. In A. Dastyari, A. Nethery, & A. Hirsch (Eds.), *Refugee externalisation policies: Responsibility, legitimacy and accountability* (pp. 171–188). Routledge.

Cappi, V., Barretta, P., Musarò, P., & Parmiggiani, P. (2021). Africa immaginata dai giovani. Il ruolo delle narrazioni mediali nella costruzione di un continente altro. *Problemi dell'informazione, 3*, 303–326. https://doi.org/10.1445/102510

Castles, S., & Miller, M. J. (1993). *The age of migration: International population movements in the modern world (1. Dr)*. Guilford Press.

Castoriadis, C. (1987). *The imaginary institution of society (1st MIT press ed)*. MIT Press.

Chakrabarty, D. (2017). The politics of climate change is more than the politics of capitalism. *Theory, Culture & Society, 34*(2–3), 25–37. https://doi.org/10.1177/0263276417690236

Chandler, D. (2014). *Resilience: The governance of complexity*. Routledge. https://doi.org/10.4324/9781315773810

Chauvin, S., & Garcés-Mascareñas, B. (2014). Becoming less illegal: Deservingness frames and undocumented migrant incorporation: Becoming less illegal. *Sociology Compass, 8*(4), 422–432. https://doi.org/10.1111/soc4.12145

Chouliaraki, L. (2013). *The ironic spectator: Solidarity in the age of post-humanitarianism*. Polity Press.

Chouliaraki, L. (2017). Symbolic bordering: The self-representation of migrants and refugees in digital news. *Popular Communication, 15*(2), 78–94. https://doi.org/10.1080/15405702.2017.1281415

Chouliaraki, L., & Georgiou, M. (2022). *The digital border: Migration, technology, power*. New York University Press.

Chouliaraki, L., & Musarò, P. (2017). The mediatized border: Technologies and affects of migrant reception in the Greek and Italian borders. *Feminist Media Studies, 17*(4), 535–549. https://doi.org/10.1080/14680777.2017.1326550

Cohen, S. (2011). *Folk devils and moral panics: The creation of the mods and rockers*. Routledge.

Colombo, E. (2011). *Le società multiculturali (2a ed)*. Carocci ed.

Couldry, N. (2010). *Why voice matters: Culture and politics after neoliberalism*. Sage.

Crawley, H., Düvell, F., Jones, K., McMahon, S., & Sigona, N. (2016). *Destination Europe?Understanding the dynamics and drivers of Mediterranean migration in 2015*. MEDMIG Final Report. http://www.medmig.info/research-brief-destination-europe.pdf

Cuttitta, P. (2015). 'Borderizing' the Island Setting and Narratives of the Lampedusa 'Border Play.' *ACME: An International Journal for Critical Geographies*, 196–219 Pages. https://doi.org/10.14288/ACME.V13I2.1004

Dal Lago, A. (2012). *Non-persone: L'esclusione dei migranti in una società globale* (Nuova ed., 5. ed). Feltrinelli.

Danowski, D., & Viveiros de Castro, E. (2017). *Esiste un mondo a venire?: Saggio dulle paure della fine*. Nottetempo.

Davoudi, S., & Machen, R. (2022). Climate imaginaries and the mattering of the medium. *Geoforum, 137*, 203–212. https://doi.org/10.1016/j.geoforum.2021.11.003

De Haas, H. (2007). Turning the tide? Why development will not stop migration. *Development and Change, 38*(5), 819–841. https://doi.org/10.1111/j.1467-7660.2007.00435.x

De Martino, E. (2023). *The end of the world: Cultural apocalypse and transcendence* (1st ed.). University of Chicago Press.

Decimo, F. (2024). Introduction: Migration, reproduction, and the place of families. In F. Decimo (Ed.), *Lives in motion* (pp. 1–29). Springer Nature Switzerland. https://doi.org/10.1007/978-3-031-65583-8_1

Derman, B. B. (2020). *Struggles for climate justice: Uneven geographies and the politics of connection*. Springer International Publishing. https://doi.org/10.1007/978-3-030-27965-3

Duffield, M. (2001). Governing the borderlands: Decoding the power of aid. *Disasters, 25*(4), 308–320. https://doi.org/10.1111/1467-7717.00180

Duffy, H., & Leppänen, K. (Eds.). (2024). *Storying the ecocatastrophe: Contemporary narratives about the environmental collapse*. Routledge, Taylor & Francis Group.

Entman, R. M. (1993). Framing: Toward clarification of a fractured paradigm. *Journal of Communication, 43*(4), 51–58. https://doi.org/10.1111/j.1460-2466.1993.tb01304.x

European Agenda for the Integration of Third-Country Nationals. (2011). *Communication from the Commission to the European Parliament, the Council, the European Economic and Social Committee and the Committee of the Regions*. European Commission. https://eur-lex.europa.eu/LexUriServ/LexUriServ.do?uri=COM:2011:0455:FIN:en:PDF

Faloppa, F. (2020). *Odio: Manuale di resistenza alla violenza delle parole* (Prima edizione). UTET.

Fargione, D., & Concilio, C. (Eds.). (2018). *Antroposcenari: Storie, paesaggi, ecologie*. Convegno Antroposcenari, storie, paesaggi, ecologie, Bologna. Società editrice Il mulino.

Fassin, D. (2011). *Humanitarian reason: A moral history of the present*. University of California Press. https://doi.org/10.1525/california/9780520271166.001.0001

Fisher, M. (2009). *Capitalist realism: Is there no alternative?* John Hunt Publishing.

Garelli, G., & Tazzioli, M. (2017). *Tunisia as a revolutionized space of migration*. Palgrave Macmillan.

Gatta, G. L., Mitsilegas, V., & Zirulia, S. (2021). *Controlling immigration through criminal law: European and comparative perspectives on "crimmigration"*. Hart.

Ghosh, A. (2016). *The great derangement: Climate change and the unthinkable* (First published in Allen Lane by Penguin Books India 2016). Allen Lane, an imprint of Penguin Books.

Giacomelli, E. (2020). La banalità del mare. Frontiere, quarantena e navi da crociera ai tempi del COVID-19. *Scritture migranti*, 155–178 Paginazione. https://doi.org/10.6092/ISSN.2035-7141/13874

Giacomelli, E. (2023). *Panicocene: Narrazioni su cambiamenti climatici, regimi di mobilità e migrazioni ambientali*. Franco Angeli.

Giacomelli, E., Musarò, P., & Parmiggiani, P. (2021). The «invisible enemy» and the usual suspects. How Covid-19 re-framed migration in Italian media representations. *Sociologia Della Comunicazione, 60*, 119–136. https://doi.org/10.3280/SC2020-060011

Giacomelli, E., & Walker, S. (2022). On board the quarantine-ship as "floating hotspot": Creeping externalization practices in the Mediterranean Sea. *Ethnography*, 146613812211005. https://doi.org/10.1177/14661381221100532

Goffman, E. (1986). *Frame analysis: An essay on the organization of experience* (Northeastern University Press ed). Northeastern University Press.

Goodman, S., Sirriyeh, A., & McMahon, S. (2017). The evolving (re)categorisations of refugees throughout the "refugee/migrant crisis". *Journal of Community & Applied Social Psychology, 27*(2), 105–114. https://doi.org/10.1002/casp.2302

Guiraudon, V., & Joppke, C. (Eds.). (2003). *Controlling a new migration world* (0th ed.). Routledge. https://doi.org/10.4324/9780203164938

Hinchman, L. P., & Hinchman, S. (Eds.). (1997). *Memory, identity, community: The idea of narrative in the human sciences*. State University of New York Press.

Hjarvard, S. (2008). The mediatization of society: A theory of the media as agents of social and cultural change. *Nordicom Review, 29*(2), 102–131. https://doi.org/10.1515/nor-2017-0181

Hodder, P., & Martin, B. (2009). Climate crisis? The politics of emergency framing. *Economic and Political Weekly, 44*(36), 55–60.
IOM. (2024). *World Migration Report 2024*. https://worldmigrationreport.iom.int/msite/wmr-2024-interactive/
Ionesco, D., Mokhnacheva, D., & Gemenne, F. (2017). *The atlas of environmental migration (first published 2017) [map]*. Routledge.
IPCC. (2022). *Climate Change 2022: Impacts, Adaptation, and Vulnerability [Contribution of Working Group II to the Sixth Assessment Report of the Intergovernmental Panel on Climate Change]*. Intergovernmental Panel on Climate Change.
ISMU. (2021). *Ventisettesimo Rapporto sulle migrazioni*, 2021. https://www.ismu.org/ventisettesimo-rapporto-sulle-migrazioni-2021/
Kermode, F. (1967). *Sense of an ending: Studies in the theory of fiction*. Oxf. UP(NY).
Kermode, F., Montefoschi, G., Giglioli, D., & Zuppet, R. (2020). *Il senso della fine: Studi sulla teoria del romanzo*. Il Saggiatore.
Klein, N. (2007). *The shock doctrine: The rise of disaster capitalism*. Allen Lane.
Krzyżanowski, M., & Forchtner, B. (2016). Theories and concepts in critical discourse studies: Facing challenges, moving beyond foundations. *Discourse & Society, 27*(3), 253–261. https://doi.org/10.1177/0957926516630900
Kunelius, R., & Roosvall, A. (2021). Media and the climate crisis. *Nordic Journal of Media Studies, 3*(1), 1–19. https://doi.org/10.2478/njms-2021-0001
Langer, A. (1994). *La conversione ecologica potrà affermarsi soltanto se apparirà socialmente desiderabile*. https://www.alexanderlanger.org/it/140/268
Latour, B. (2020). *Tracciare la rotta. Come orientarsi in politica*. Raffaello Cortina Editore.
Levantesi, S. (2021). *I bugiardi del clima: Potere, politica, psicologia di chi nega la crisi del secolo (Prima edizione)*. GLF editori Laterza.
Lomborg, B. (2001). *The skeptical environmentalist: Measuring the real state of the world*. Cambridge University Press.
MacGregor, S. (2009). A stranger silence still: The need for feminist social research on climate change. *The Sociological Review, 57*(2_suppl), 124–140. https://doi.org/10.1111/j.1467-954X.2010.01889.x
Malkki, L. H. (1996). Speechless emissaries: Refugees, humanitarianism, and Dehistoricization. *Cultural Anthropology, 11*(3), 377–404. https://doi.org/10.1525/can.1996.11.3.02a00050
Malvestio, M. (2021). *Raccontare la fine del mondo: Fantascienza e antropocene*. Nottetempo.
Mbembe, A. (2020). *Brutalisme*. La Découverte.
McCombs, M. (2005). A look at agenda-setting: Past, present and future. *Journalism Studies, 6*(4), 543–557. https://doi.org/10.1080/14616700500250438

Meschiari, M. (2019). *La grande estinzione: Immaginare ai tempi del collasso*. Armillaria.

Mezzadra, S., & Neilson, B. (2013). *Border as method, or, the multiplication of labor*. Duke University Press.

Milkoreit, M. (2017). Imaginary politics: Climate change and making the future. *Elementa: Science of the Anthropocene, 5*, 62. https://doi.org/10.1525/elementa.249

Molnar, P. (2022). Territorial and digital Borders and migrant vulnerability under a pandemic crisis. In A. Triandafyllidou (Ed.), *Migration and pandemics* (pp. 45–64). Springer International Publishing. https://doi.org/10.1007/978-3-030-81210-2_3

Monbiot, G. (2022). 5.11 changing the media narrative. In G. Thunberg (Ed.), *The climate book* (pp. 308–215). Penguin Books.

Morton, T. (2012). *The ecological thought* (1. Harvard University Press paperback ed). Harvard University Press.

Musarò, P. (2017). Mare nostrum: The visual politics of a military-humanitarian operation in the Mediterranean Sea. *Media, Culture & Society, 39*(1), 11–28. https://doi.org/10.1177/0163443716672296

Musarò, P., & Parmiggiani, P. (Eds.). (2014). *Media e migrazioni: Etica, estetica e politica del discorso umanitario*. FrancoAngeli.

Musarò, P., & Parmiggiani, P. (2022). *Ospitalità mediatica. Le migrazioni nel discorso pubblico*. FRANCOANGELI.

Nail, T. (2016). *Theory of the border*. Oxford university press.

Nixon, R. (2013). *Slow violence and the environmentalism of the poor* (First Harvard University Press paperback edition). Harvard University Press.

Omizzolo, M. (2019). *Essere migranti in Italia: Per una sociologia dell'accoglienza*. Meltemi.

Paasi, A. (1996). *Territories, boundaries, and consciousness: The changing geographies of the Finnish-Russian border*. Wiley.

Papastergiadis, N. (2012). *Cosmopolitanism and culture*. Polity Press.

Park, R. E., & Burgess, E. W. (2014). *Introduction to the science of sociology*. CreateSpace Independent Publishing Platform.

Pezzani, L., & Heller, C. (2019). AIS politics: The contested use of vessel tracking at the EU'S maritime frontier. *Science, Technology, & Human Values, 44*(5), 881–899. https://doi.org/10.1177/0162243919852672

Pinelli, B. (2019). *Migranti e rifugiate: Antropologia, genere e politica (Prima edizione)*. Edizioni Libreria Cortina.

Rancière, J. (2009). *The politics of aesthetics: The distribution of the sensible* (reprinted). *Continuum*.

Rath, J., & Schutjens, V. (2019). Advancing the Frontiers in ethnic entrepreneurship studies. *Tijdschrift voor Economische en Sociale Geografie, 110*(5), 579–587. https://doi.org/10.1111/tesg.12398

Ribeiro, S., & Soromenho-Marques, V. (2022). The techno-optimists of climate change: Science communication or Technowashing? *Societies, 12*(2), 64. https://doi.org/10.3390/soc12020064

Rorty, R. (1989). *Contingency, irony, and solidarity*. Cambridge University Press.

Said, E. W. (1979). *Orientalism*. Vintages books.

Sandel, M. J. (2020). *The tyranny of merit: What's become of the common good?* (1st ed.). Farrar, Straus and Giroux.

Saramago, J. (2021). *Cecità* (18th ed.). Feltrinelli.

Schinaia, C. (2025). *Against catastrophism: Climate change, pandemics, and Hope for the future* (1st ed.). Taylor & Francis Group.

Schuster, J. (2014). How to write the disaster. *The Minnesota Review, 2014*(83), 163–171. https://doi.org/10.1215/00265667-2782363

Scranton, R. (2015). *Learning to die in the Anthropocene: Reflections on the end of a civilization*. City Lights Books.

Shapiro, M. J. (1988). *The politics of representation: Writing practices in biography, photography, and policy analysis*. University of Wisconsin Press.

Sheller, M. (2018). *Mobility justice: The politics of movement in the age of extremes*. Verso.

Silvestrini, G., & Zorzoli, G. B. (2020). *Le trappole del clima: E come evitarle*. Ambiente.

Taylor, C. (2003). *Modern social imaginaries (4. print)*. Duke Univ. Press.

Tazzioli, M. (2020). *The making of migration: The biopolitics of mobility at Europe's Borders*. SAGE Publications. https://doi.org/10.4135/9781526492920

Thunberg, G. (2022). *The climate book*. Penguin Press.

Van Aken, M. (2020). *Campati per aria*. Elèuthera.

Van Delft, H., Gorter, C., & Nijkamp, P. (2000). In search of ethnic entrepreneurship opportunities in the City: A comparative policy study. *Environment and Planning C: Government and Policy, 18*(4), 429–451. https://doi.org/10.1068/c9873

Vanhala, L. (2016). *Strategic ambiguity: How loss and damage became a part of global climate policy. New security beat*. Environmental Change and Security Program (ECSP). https://www.newsecuritybeat.org/2016/11/strategic-ambiguity-loss-damagepart-global-climate-policy-cop-22/

Vaughan-Williams, N. (2015). *Europe's border crisis: Biopolitical security and beyond*. Oxford University Press. https://doi.org/10.1093/acprof:oso/9780198747024.001.0001

Walters, W. (2011). Foucault and frontiers: Notes on the birth of the humanitarian border. In U. Bröckling, S. Krasmann, & T. Lemke (Eds.), *Governmentality: Current issues and future challenges* (pp. 138–164). Routledge.

Walters, W., Heller, C., & Pezzani, L. (Eds.). (2022). *Viapolitics: Borders, migration, and the power of locomotion*. Duke University Press.

Watson, S. (2011). The 'human' as referent object?: Humanitarianism as securitization. *Security Dialogue, 42*(1), 3–20. https://doi.org/10.1177/0967010610393549

Williams, R. (1976). *Keywords: A vocabulary of culture and society*. Fourth Estate.

Wodak, R. (2015). *The politics of fear: What right-wing populist discourses mean*. Sage.

Wu Ming Collective. (2013). *Storie #notav. Un anno e mezzo nella vita di Marco Bruno*. http://www.wumingfoundation.com/giap/2013/07/storie-notav-un-anno-e-mezzo-nellavita-di-marco-bruno/

Xausa, C. (2020). Cambiamento climatico, genere e intersezionalità: Narrazioni r-esistenti alla climate fiction apocalittica. In *Genere e R-esistenze in movimento: Soggettività, azioni e prospettive (a cusa di Coppola, M. M.; Donà, A.; Poggio, B.; Tuselli, A.)*. (pp. 99–108). Università di Trento.

Yamamura, S., & Lassalle, P. (2020). Proximities and the emergence of regional industry: Evidence of the liability of smallness in Malta. *European Planning Studies, 28*(2), 380–399. https://doi.org/10.1080/09654313.2019.1668915

Yuen, E. (2012). The politics of failure have failed: The environmental movement and catastrophism. In S. Lilley, D. McNally, E. Yuen, & J. Davis (Eds.), *Catastrophism: The apocalyptic politics of collapse and rebirth*. PM Press.

Yusoff, K. (2018). *A billion black Anthropocenes or none*. University of Minnesota Press.

Yusoff, K., & Gabrys, J. (2011). Climate change and the imagination. *WIREs Climate Change, 2*(4), 516–534. https://doi.org/10.1002/wcc.117

Yuval-Davis, N., Wemyss, G., & Cassidy, K. (2018). Everyday bordering, belonging and the reorientation of British immigration legislation. *Sociology, 52*(2), 228–244. https://doi.org/10.1177/0038038517702599

Zaborowski, R., & Georgiou, M. (2019). Gamers versus zombies? Visual mediation of the citizen/non-citizen encounter in Europe's 'refugee crisis'. *Popular Communication, 17*(2), 92–108. https://doi.org/10.1080/15405702.2019.1572150

Zanfrini, L. (2016). *Introduzione alla sociologia delle migrazioni (Prima edizione)*. GLF editori Laterza.

Zolberg, A. R. (1997). Richiesti ma non benvenuti. *Rassegna Italiana di Sociologia, 1*, 19–40. https://doi.org/10.1423/2445

CHAPTER 4

Climate (Im)Mobilities

Introduction

For centuries, human mobility has been shaped by environmental changes. Seasonal movements along established routes were not merely survival strategies but integral to societal structures. The *Atlas of Environmental Migration* (Ionesco et al., 2017) traces these patterns back 45,000 years, underscoring how climate fluctuations and natural disasters have historically influenced population distribution. However, it is only in recent decades that the international community has explicitly acknowledged the deep interconnection between climate change and human (im)mobility (Laczko & Aghazarm, 2009). As Piguet (2022) highlights, in the early twenty-first century, the migration-environment nexus has garnered growing scholarly attention, reflecting the salience and urgency of this issue.

While scholarly, journalistic, and activist engagement with climate-induced human mobility continues to grow, persistent conceptual ambiguities pervade discussions around narrative framing and justice advocacy in this domain. The emerging field of climate (im)mobilities research demonstrates dynamic interdisciplinary vitality, drawing from environmental science, migration studies, and forced displacement literatures while also transcending these disciplinary boundaries. This nascent paradigm recognizes that climate (im)mobilities constitute an emergent phenomenon, not simply a sum of climate change impacts and

E. Giacomelli and P. Musarò, *Climate Mobility Justice*,
https://doi.org/10.1007/978-3-032-07907-7_4

(im)mobility patterns. Following the principle of non-additive complexity (where $1 + 1 \neq 2$ but instead generates a third ontological category), climate (im)mobilities must be analyzed as a distinct socio-ecological phenomenon, characterized by unique causal pathways, feedback loops, and governance challenges that cannot be reduced to its constituent elements. The recent conceptual evolution toward understanding climate (im)mobilities as emergent socio-ecological systems (Boas et al., 2019) demands parallel innovations in justice-oriented narrative construction. This shift challenges reductionist causal models while acknowledging the irreducible complexity of lived experiences in the context of environmental change. Hence, the nexus between climate change and human (im)mobility represents one of the most pressing challenges of our time—demanding a multifaceted approach to justice that transcends conventional analytical boundaries.

In the liminal space of the Anthropocene, our journey begins at the confluence of the physical and the metaphysical, the present and the future, where the tangible impacts of rising seas and a scorched earth intertwine with the intangible realms of human experience, emotion, and imagination. We posit that to truly comprehend the depth and breadth of climate (im)mobilities, one must adopt a perspective that is at once panoramic and microscopic, capable of discerning both the grand patterns of global systems and the minute details of individual lives disrupted.

Our analysis begins by situating climate (im)mobilities within the broader context of the Anthropocene, critically examining how this concept both illuminates and obscures the root causes of our current ecological crisis. We then advocate a shift toward the notion of the Capitalocene, which more accurately captures the systemic inequalities and historical injustices that underpin climate-induced displacement. This reframing enables a more incisive interrogation of the power dynamics shaping climate (im)mobilities and the Panicocene narratives surrounding them.

As we traverse this intellectual terrain, we shall encounter justice in its myriad forms—a shape—shifting concept that defies simple categorization. We will explore its scalar nature, from the intimate sphere of local communities to the vast expanses of international relations, and from earthbound realities to a form of cosmopolitanism that—as Papastergiadis (2023, p. 184) writes—'needs to be seen from both the polis and the cosmos.' In doing so, we aim to construct a holistic framework that captures the full spectrum of justice in the context of climate (im)mobilities. These different facets of justice, we argue, are fundamentally interconnected and must be considered historically and holistically to effectively address the complexities of climate (im)mobilities. As Carmen G. Gonzalez (2021, p. 114) puts it, 'The extraction of wealth from nature was operationalized through slavery, the conquest and dispossession of indigenous peoples, and the colonial and postcolonial domination by white Euro-American elites of large segments of the world's population.'

Recognizing climate (im)mobility as a justice issue requires moving beyond securitized narratives toward more inclusive and transformative solutions. Central to our analysis is the dichotomy between competing security narratives. On one hand, we witness the entrenchment of a fortress mentality, materialized in the proliferation of militarized borders and surveillance regimes, often justified through apocalyptic projections of 'climate conflict' (Hartmann, 2010). On the other hand, alternative imaginaries emerge that are grounded in principles of care, solidarity, cosmopolitan ideas, and ecological restoration. This discursive shift is exemplified by Vince's (2022) vision of a 'Nomad Century' which reframes mobility not as a civilization threat but as a form of adaptive genius. This tension, we contend, is not merely narrative nor academic—it is profoundly consequential, shaping policies, lived experiences, and the very fabric of our global society. We also examine how these contrasting narratives shape public perceptions of climate (im)mobilities often work to the detriment of those most vulnerable to climate change impacts.

Media narratives play a pivotal role in shaping public perceptions of climate (im)mobilities. Sensationalist reporting often portrays migration as a mass South-to-North movement, reinforcing securitized and racialized fears. This discursive battleground reveals deeper epistemological shifts: a contest between crisis frameworks—often used to justify securitization measures—and rights-based paradigms that assert the fundamental right to mobility with dignity (Methmann & Oels, 2015). Yet, these narratives frequently overlook instances of internal displacement in the Global North caused by extreme weather events, such as the floods in Emilia-Romagna (Italy, 2023) (Fig. 4.1) and Valencia (Spain, 2024). Similarly, media attention often centers on small Pacific island nations,[1] such as Tuvalu and Kiribati, while neglecting broader and more complex patterns of climate-induced (im)mobilities.

[1] https://www.policyforum.net/new-zealand-a-global-leader-on-climate-and-displacement/

Fig 4.1 The floods of the Senio river in Cotignola (Italy), September 19, 2024. Ph. Michele Lapini. Courtesy of the photographer Michele Lapini

As emphasized in previous chapters, the framing and narration of climate change and migration within media and policy discourses carry substantial justice implications. Climate change-induced (im)mobility is an increasingly urgent and controversial topic; yet reductive approaches risk sensationalizing it, often perpetuating stereotypes and simplistic interpretations. To better capture the pervasive fear and uncertainty surrounding climate (im)mobilities, we introduce the concept of Panicocene—a double crisis narrative rooted in the panic over mass displacement caused by environmental disasters.

In line with Ranciere's framework, we underscore throughout the chapter the importance of sensory experiences, emotions, and imaginaries in shaping our understanding of—and response to—climate (im)mobilities. We argue that these subjective elements play a crucial role in how different actors perceive, interpret, and engage with the issue. Critiquing the dominant portrayal of climate migration as a security threat that fosters militarized responses as opposed to policies centered on justice and human rights, we call for the de-securitization and de-militarization of climate migration narratives. In their place, we advocate for alternative frameworks such as mobility justice (Sheller, 2020) and the coloniality of power (Quijano, 2000). Finally, we identify emergent narrative elements that will be examined more closely in Chapters 5 and 6 through specific case studies.

Migration and Climate Change: A Contested Nexus

The nexus between climate change and human mobility presents a complex and multifaceted challenge that defies simplistic causal attributions. Migration decision-making unfolds through iterative processes shaped by intersecting socioeconomic, political, and environmental factors, rather than discrete events (de Sherbinin et al., 2022). The factors influencing the decision-making process on migration choice are complex and multi-causal; in many cases, migration is better understood as a process than as a singular decision or event (Kelman, 2020). This complexity challenges dominant analytical frameworks that posit environmental factors as direct migration drivers, often overlooking both structural inequalities and human agency. Distinguishing environmental migration from other forms of displacement is particularly difficult, as multiple overlapping dynamics influence (im)mobility (Boas et al., 2019, p. 902).

Debates over definitions and conceptual frameworks abound (McLeman, 2018). The International Organization for Migration (IOM) defines climate migration as 'the movement of a person or groups of persons who, predominantly for reasons of sudden or progressive change in the environment due to climate change, are obliged to leave their habitual place of residence or choose to do so, either temporarily or permanently, within a State or across an international border' (IOM, 2019, p. 31). Similarly, the Intergovernmental Panel on Climate Change (IPCC) has long acknowledged climate change as a driver of large-scale human displacement (IPCC, 1990). This challenge is further compounded by the ongoing debate over how to classify individuals relocating from areas impacted by climate change—whether as refugees, displaced persons, or migrants—while disentangling the manifold drivers of migration processes.

The 'environmental refugee' first appeared in Hinnawi's (1985) United Nations Environment Programme (UNEP) report entitled *Environmental Refugees*. However, this, or the more common category of 'climate refugee,' is now widely questioned. Given the multi-causal nature of most migration, the IPCC has itself acknowledged the difficulty of identifying any individual as a clear-cut climate migrant and that 'there is broad agreement in the scientific and legal literature that the use of the term climate refugee is scientifically and legally problematic' (IPCC, 2014, p. 771).

The Nansen Initiative (2015) and its successor, the Platform on Disaster Displacement, have played a central role in raising global awareness of disaster-related mobility. These state-led processes have fostered nuanced research on the phenomenon, leading to the integration of disaster and climate change considerations into key international frameworks, such as the Sendai Framework for Disaster Risk Reduction (2015–2030), the 2030 Agenda for Sustainable Development, and the Agenda for Humanity. Nonetheless, the 1951 Refugee Convention does not account for environmental causes of displacement, leaving critical protection gaps.

The 2018 *Global Migration Compact,* on the other hand, offers a promising framework by addressing mobility in the context of disasters and climate change. It emphasizes information sharing, resilience-building strategies, integration of displacement into disaster preparedness, and expanded migration pathways. Although non-binding, the Global *Migration Compact* is seen as a potential foundation for future policy advancements. Additionally, the Paris Agreement (2015) indirectly acknowledges climate-induced mobility by linking climate action to human rights. The UNFCCC's Task Force on Displacement (2017) further aims to address displacement caused by climate impacts. Yet despite these efforts, high-level international fora—such as COP26 (2021)—failed to explicitly address climate-induced migration, underscoring ongoing challenges in achieving global consensus around this issue.

Put plainly, many states resist recognizing climate-related displacement because doing so would entail new obligations regarding reception and protection—obligations they are disinclined to assume (Scissa, 2024). This political reluctance is often rationalized by pointing to the absence of scientific models capable of reliably predicting the number of individuals who may migrate in the future due to climate change and environmental degradation. Without such precise data, states fear that explicitly recognizing this connection and the associated need for protection could open the door to large numbers of climate refugees. In this sense, current legal ambiguities reproduce the tension between the national and cosmopolitan scales of justice already embedded in the Geneva Refugee Convention, while legitimating policies of externalization and the erosion of asylum rights.

Securitization is key here. As Rychnovskà (2014) notes, the construction of threats is central to security discourse—and the media, as Bleich et al. (2015) and Watson (2011) argue, often act as securitizing agents themselves, especially in their portrayals of migrants and refugees. Several scholars have warned against an alarmist narrative of impending mass climate migration, which is often instrumentalized to justify the securitization of human mobility. Renou and Diallo (2019), for instance, decry how divergences over terminology and methodology, combined with uncertainties about the future development of climate change (projections), an alarmist media logic, and the increasing invocation of 'climate migration' as a bogeyman, leads to intensified political weaponization and the erection of (more) walls and security measures.

Uncertainties are especially pronounced when examining slow-onset environmental processes—what Nixon (2013) calls 'slow violence'—and their role in shaping mobility. These slow transformations complicate attempts to determine who will migrate, where (within the same country, in neighboring countries, or internationally), and for how long (Flavell et al., 2020).

While displacements dictated by slow-moving environmental changes are difficult both to define and to mediate, data on displacement from sudden environmental upheavals such as earthquakes, cyclones, hurricanes, and typhoons are more readily available. The Internal Displacement Monitoring Centre (IDMC) provides annual estimates of the number of people displaced within the national state borders due to these more punctual disasters. Yet even these figures are inaccurate and incomplete: they indicate the scale of displacement but not its duration, destinations, or qualitative impact—such as the situational severity—on displaced populations. Nor do they account for the vulnerability (present and future) of people who are forced to migrate (Flavell et al., 2020).

We can find catastrophic predictions about the number of people displaced by climate change in the World Bank's recent *Groundswell* report (Clement et al., 2021), which projected that by 2050, up to 216 million people may be forced to leave their homes. It is important to note that this figure refers to internal migration and not cross-border migration, which is how environmental migration is often narrated and represented. However, alarmist narratives from the perspective of the

Global North continue to the depiction of migrants as incoming 'threats' from the Global South, when, in reality, the vast majority of movements occur within national or regional borders.

Moreover, not all vulnerable people can move. Migration requires resources—social and economic capital—and some of the most affected by environmental change are also the least mobile. Yet such nuances are often lost or flattened in media discourses, which (mis)use numbers to exacerbate fear narratives (Durand-Delacre et al., 2021; Lietaer & Durand-Delacre, 2021). As Baldwin and Bettini (2017, p. 1) argue, climate change itself is portrayed as 'a migration crisis in the making.'

To respond to these limitations of climate migration research, we draw from the new mobilities paradigm (Sheller & Urry, 2006), which interrogates the organization of power around systems of governing mobility, immobility, channels, and borders across various scales. This theoretical lens highlights how climate migration research has tended to under-theorize climate change's impacts on existing mobility patterns and the recursive relationship between mobility practices and climate change (Baldwin et al., 2019). Thus, in reflecting on climate mobility justice, we join other authors in promoting a shift from the narrow concept of 'climate migration' to a broader focus on 'climate (im)mobilities' (Boas et al., 2018; Parsons, 2019). This move allows for a more multi-layered analysis of how climate change impacts—and is impacted by—the (im)mobility practices of communities and individuals. Climate mobility can thus be visualized along a continuum: from forced displacement and planned relocation to voluntary mobility or immobility.

Anthropocene Mobilities

Fears over climate migration have fueled the false notion that the climate emergency is, first and foremost, a border emergency. Instead, as Neef and Sheller argue, 'we can reframe Anthropocene mobilities as a broader problem of energy over-consumption within capitalist extractive economies, for which a 'climate debt' is owed' (Neef & Sheller, 2024, p. 649).

In contrast to conventional policies centered on border closures and the expulsion of so-called 'illegal migrants,' the climate (im)mobility justice framework advocates for a multidimensional policy approach that recognizes climate-induced migration as a legitimate adaptive response—one that warrants protection under international law rather than suppression and criminalization. Achieving this requires a reconceptualization of those displaced by climate change, understanding them not as stateless claimants but as displacees within a globalized network of intersecting mobility regimes driven by fossil fuel extraction (Baldwin et al., 2019, p. 291). In this sense, they are not merely rights-bearers tied to state-membership but citizens of the cosmos, and, as such, free to move across the planet.

To better understand and respond to what Andrew Baldwin et al. (2019) called 'Anthropocene mobilities,' it is crucial to interrogate the concept of the Anthropocene itself. Originally coined by Nobel Prize-winning chemist Paul Crutzen to describe the current geological epoch in which human activities have fundamentally altered the Earth's systems, the term has become widely used in scientific and cultural discourse. Yet it has also drawn sharp criticism: for many thinkers, not only does it obscure disparities in responsibility and impact, it also does not name the Western cultural model the capitalist system at the root of planetary degradation. While the term Anthropocene certainly reflects human responsibility and interconnectedness in connection to our changing planet, its effectiveness depends on whether it is framed as an inevitability or as a call for transformative action. Recognizing the cognitive impact of this framing is crucial for fostering meaningful engagement with the climate crisis.

As Luigi Pellizzoni (2021) points out, the Anthropocene is the name of a governmental device that lends a performative force to the crisis by framing climate change as both cause and effect and by presenting its consequences as an inescapable destiny to which humanity can only adapt. As it shifts veridical force from diagnosis to prognosis and technocratic therapy, the Anthropocene provides us with a sense of reality and human-planetary fate produced by the upheaval of techno-scientific, political, and economic modernity—that of capitalism as an institutionalized social order—to which we must simply adjust. This narrative bolsters the current production system by promising, at best, 'ecologically sustainable' control by elites who, with the aim of making certain places on Earth more 'sustainable,' generate oases of privilege protected by walls and immune systems, real and virtual (Manghi, 2004).

Multiple critics (Van Aken, 2020; Barca, 2020; Solórzano and Yosso 2002) denounce how dominant Anthropocene narratives often trivialize and mystify reality by obscuring the systemic intersection of racism, colonialism, class inequality, heteropatriarchy, and human supremacy in the production of the environmental crisis (Armiero, 2021). The social imaginary linked to the Anthropocene leads to a general lack of accountability as well as the depoliticization and gradual normalization of the crisis (Giuliani, 2021), effectively perpetuating the social and economic injustices baked into the climate crisis. It is an imaginary built over at least half a century, in which symptoms (environmental disasters) and not causes (political, economic, and social) take center stage.

Seen from the perspective of racialized subjectivities, the so-called climate emergency is hardly new. As Stacy Alaimo (2010) observes, environmental racism in the United States, but also elsewhere, has long determined the placement of toxic waste release sites in communities of color. Desertification, disruption of ecosystems, reduction of environments suitable for life, and increased exposure and physical vulnerability have always been part of what colonial capitalism has meant for the Global South (Povinelli, 2021). Kathryn Yusoff (2018, p. 105) calls this erasure the 'White Anthropocene,' a Western geological imaginary of the 'ecological crisis' that renders invisible the histories of non-Whites and their colonial experiences. A colonial arrogance persists on the part of present-day 'collapsologists' when they talk about a new collapse while concealing the connections that exist to modern colonization, slavery, racism, the genocides of indigenous peoples, and the destruction of their environments (Servigne & Stevens, 2021).

The Anthropocene is thus not an event, but an ongoing condition: it is the lived reality of colonial capitalism since its inception. What we usually think of as a novel event (the imminent climate catastrophe) is in fact a daily reoccurrence for others. Having already happened many times over, for the long-colonized it is not a disaster to come but an ancestral catastrophe that is not the same kind of thing-event as the impending one, nor does it operate within the same temporality. As Povinelli writes: 'When we begin with the catastrophe of colonialism and enslavement, the location of contemporary climatic, environmental, and social collapse rotates and mutates into something else entirely. Ancestral catastrophes are past and present; they keep arriving out of the ground of colonialism and racism rather than emerging over the horizon of liberal progress' (2021, p. 3).

A range of alternative terms have been proposed to critique the current geological epoch: Wasteocene, Plantationocene, Negrocene,

Chtulucene, and others. Among these, Capitalocene most directly challenges the notion of 'anthropogenic' climate change as a generalized human phenomenon, anthropos, insisting instead on its 'capitalogenic' origins (Moore, 2017, p. 29). Unlike the Anthropocene, which attributes environmental degradation to an undifferentiated 'humanity,' the Capitalocene highlights how capitalist expansion, fossil fuel dependence, and colonial histories have disproportionately contributed to climate crises, and shaped patterns of displacement and mobility. Taken up in recent years by Andreas Malm, Jason Moore, Donna Haraway, and others, the concept shifts analytical focus from humanity as a whole to the political economy of corporate globalization and has the advantage of naming the culprit of climate change, thereby mustering greater political traction (Haraway, 2015; Hamilton et al., 2015; Parenti & Moore, 2016). Identifying the economic nature of our geological present, Andreas Malm (2016) argues, 'ours is not the geological epoch of humanity but of capital.'

In other words, it is not humanity in general that determines our direction, but rather the economics of petro capitalism in the form of lobbying, greenwashing, climate change denial, media spectacularization, and obfuscation (Demos, 2016). Françoise Vergès (2017) extends this critique with the term 'racial capitalocene' emphasizing how colonialism, slavery, and 'the global use of the color line' have led to the devaluation of both human and non-human life.

In understanding contemporary environmental crises, it is crucial to remain attuned to the ways in which 'destruction in the colonial era becomes visible in the post-colonial era' (Vergès 2017, p. 77). The concept of the Capitalocene—which situates climate change within the historical and structural dynamics of capitalism—offers a critical lens for understanding climate (im)mobilities. Climate-induced (im)mobilities, therefore, are not merely an outcome of environmental change but a manifestation of socioeconomic and political systems that have historically produced discrimination, exclusion, and vulnerabilities while enabling mobility for some and restricting it for others. Capitalist-driven resource extraction, land dispossession, and exploitative labor markets have forced marginalized communities into precarious conditions, rendering them more susceptible to climate-induced displacement. Simultaneously, restrictive border regimes and securitized migration policies criminalize the movement of the displaced, reinforcing global hierarchies of mobility and immobility. Thus, analyzing climate (im)mobilities through the Capitalocene framework underscores the need for policies that address the structural inequalities and injustices underpinning climate displacement in contrast to those simply managing migration as a controllable crisis.

How, then, to explain the current global inaction on climate (im)mobility injustice? Naomi Klein offers one answer in *This Changes Everything*: 'We are stuck because the actions that would give us the best chance of averting catastrophe—and would benefit the majority—are extremely threatening to an elite minority that has a stranglehold over our economy, our political process, and most of our majority media outlets' (Klein, 2015, p. 18).

Analyzing the justice framework through the power of narratives and representations is thus critical to fully understanding the climate mobility regime. The nexus between climate change and migration affects both the present and the future in many ways, especially as its governance now relies on assumed risk perceptions and uncertainty. As Baldwin and Bettini (2017) argue, the relationship between climate change and migration must be understood primarily as a 'power relationship,' rather than as a fact waiting to be discovered or an empirically observable phenomenon. Climate change designates a real crisis in the contemporary political economy, but its migratory effects are narrated and always mediated by all sorts of other social, cultural, political, and even geological relations. Consequently, as the two authors argue, 'the very act of thinking (or speaking or imagining) climate change as a problem of migration is explicitly political. By bending our view of the world, such thinking becomes a powerful means by which social attitudes are governed, which is what we mean when we say that climate change and human migration is a relation of power' (Baldwin & Bettini, 2017, p. 2).

This power relation is nowhere more evident than in Panicocene narratives—apocalyptic visions of climate collapse intertwined with fantasies of mass migration. These narratives generate what we might call an 'emergency squared,' multiplying fear while concealing responsibility. Although climate-induced mobilities are diverse, context-dependent, and shaped, at least in part, by existing relations of power and inequality, the Panicocene narratives highlight how these fears, often amplified by media and political rhetoric, animate our collective consciousness and our responses to the crises we face.

Climate (Im)Mobilities and Justice

Climate change is intensifying contemporary displacement to unprecedented levels, provoking new climate (im)mobilities that are shaped by existing inequalities related to gender, ethnicity, class, or resource scarcity. As Yumagulova et al. argue: 'Although significant, the existing climate mobility regime is very narrow in scope and does not adequately provide for mobility justice in terms of those people who are engaged in efforts to actively resist climate displacement (choosing to be climate immobile) through taking actions to reduce their vulnerability and adapt in place' (2023, p. 7). It is therefore essential to understand and recognize the multiplicity of climate-induced displacements and (im)mobilities.

A broadened framework of climate (im)mobility regimes pictured as 'a fragile entanglement of physical movement, representations, and practices' (Cresswell, 2010, p. 18), allows us to recognize what Everuss 2023, p. 28) defines as the 'ontological significance of climate (im)mobilities,' which exposes the falsehood of the ostensibly static pre-climate change subject. Drawing on the example of the low-lying island nation Tuvalu, Everuss explains that internal and external mobility have long been integral to the Tuvaluan people's mobile, seafaring, and oceanic historical identities. Thus, if we frame their contemporary mobilities as part of long-standing cultural traditions, 'the risk of climate change is not that supposedly static Tuvaluans will be forced to engage in movement, but instead that the culturally significant physical and imaginative mobilities of Tuvaluan people will be diminished as they will no longer be able to include travel to, from, and across the Tuvaluan Islands' (2023, p. 29).

When the effects of climate change render a location nearly uninhabitable, and individuals relocate or migrate, even temporarily, human rights and justice frameworks may provide mechanisms to alleviate human suffering and address inequalities (Ahsan, 2019; Mayer & Cournil, 2016; Naser et al., 2019).

More broadly, attending to the ontological, political, and interscalar consequences of climate (im)mobilities helps illuminate justice-related concerns—whether related to causes, lived expression, or institutional handling—and provides critical insights for reframing narratives and policies. As Farbotko et al. highlight, a climate mobility justice lens requires us 'to pay more attention to the discourses and agency of people moving or living in climate vulnerable areas to ensure that justice is achieved and human rights are respected in the climate mobility context' (2022b, pp. 13–14). Yet despite broad agreement on the need to reframe climate (im)mobility in terms of justice and human rights, it remains difficult to define precisely what these injustices entail and how best to redress them. This challenge reflects the entangled influences of agency, culture, power, temporality, and governance structures in shaping both the experience of living in and departing from areas impacted by climate change.

Moreover, notions of mobility, habitability, identity, value, and place are often perceived and contested differently by residents, institutions, and other stakeholders (Jessee, 2022; Neu & Fünfgeld, 2022; Nyantakyi-Frimpong & Dinko, 2022). These complexities may intensify as environmental conditions deteriorate, even as debates continue over how habitability is defined and who gets to define it (Duvat et al., 2021; Horton et al., 2021). Power dynamics significantly influence adaptation strategies concerning mobility (Aidoo, 2021; Bayrak et al., 2022; Paprocki, 2020), raising questions about whose perspectives are genuinely considered in the development of climate mobility solutions (Arnall et al., 2019).

As Farbotko observes: 'Climate migration knowledge founded on the concept of climate displacement risk frequently presents only unidirectional movement away from climate-vulnerable sites' (Farbotko, 2022, p. 4). With the climate migration agenda emphasizing ways to protect people who are displaced by climate change, we often lose some nuance when speaking about climate (im)mobility regimes. Let us consider two illustrative cases. First, where colonial discourses have historically denigrated certain forms of mobility—such as nomadic Indigenous or Roma cultures—as signs of primitiveness or underdevelopment, Indigenous scholars and activists are reclaiming them as a marker of resilience, connectivity, and kinship. For example, Yumagulova et al. (2023) explore the intersection of climate mobility, justice, and Indigenous self-determination. Drawing on case studies from First Nations in Canada, Māori in Aotearoa New Zealand, and Dayak communities in Indonesia, the authors critique dominant climate displacement narratives that marginalize Indigenous agency, mobility traditions, and governance systems.

Second, mainstream narratives around climate migration tend to fixate on vulnerable people fleeing from the Global South to the Global North while ignoring the movements of wealthy Westerners which may be based on the desire for a better climate. In their recent book *Vertical Migrations. Will the Mountains Save Us?*, Membretti et al. (2024) examine the phenomenon of 'new highlanders'—affluent urban dwellers relocating to mountain regions in search of more livable environments. Rich people are migrating, therefore, upward. On the one hand, Membretti, Barbera, and Tartari write, the COVID-19 pandemic had already shed a different kind of light on the mountains due to the housing and work solutions put in place (such as smart-working and multi-local living arrangements), but also because of a journalistic narrative that had romanticized and somewhat magnified the numbers of the mobility phenomenon. Likewise, ongoing heat waves and related extreme phenomena in the large cities of Southern Europe, for instance, have generated widespread anxiety over climate change, providing another reason for people to 'migrate to higher altitudes.' (Fig. 4.2).

Looking at data from northern Italy and Switzerland, the researchers specify that these actors are, for the moment, mostly affluent people moving to their second homes for a few months a year. It represents, however, a growing trend, partly because of narratives in which the common denominator is capital 'N' Nature—almost a mythological character or place offering regeneration. This idea of nature as therapy, though around since the 1970s, is recast today through practices such as 'forest bathing' or yoga in the woods. Climate change, in other words, is pushing mountains to serve a compensatory function vis-à-vis the city. Yet the mountains, the researchers continue, are a resource that should be accessible to the groups that need it most, such as the elderly and children, or those suffering from respiratory diseases. In about 20 years, large segments of the city-dwelling population will experience serious health risks for four months out of the year. Ensuring equitable access, they argue, will require targeted policy interventions.

Climate (im)mobility justice thus demands attention not only to differing scales of justice, but also to multiple mobilities and counter-mobilities (conceptualized as movements that reconfigure, subvert, or defy hegemonic mobility regimes). In this light, advancing justice within the realm of climate mobility, or integrating human rights considerations, proves to be a challenging endeavor indeed (Nash, 2018). Nevertheless, recognizing climate mobility as an issue of justice that moves from the polis to the cosmos is crucial for determining the most effective responses to it. We may begin by approaching climate mobility and its relationship with justice from different vantages.

A first perspective might be to 'de-naturalize' the climate crisis—which is often portrayed as an apolitical natural disaster—by applying the

Fig 4.2 Atlas of the new world. What will the atlas studied by future generations look like? How can a photograph capture the future? These questions inspired the creation of atlas of the new world. While photography typically captures the present, this project envisions the future by projecting images of possible transformations directly onto existing landscapes using a specialized slide projector. The resulting photographs are intricate and dreamlike, serving as metaphorical glimpses into a near future. In the picture: The Mont Blanc massif, once considered the symbol of perennial glaciers, has now become the symbol of climate change. Scientists predict that by 2100 the glacier mass will be only 5% of what it is today, disrupting ecosystems and the communities living there. Courtesy of the photographers Giulia Piermartiri and Edoardo Delille

mobility justice lens (Sheller, 2018). Only by adopting a more holistic notion of justice can we reveal hidden power dynamics and forms of exclusion, linking the right to move and the right to remain in a livable environment to broader questions of inequality (Giacomelli et al. 2023). A 'more integrated and conceptually sensitive approach to environmental issues' (Rose, 2012, p. 2) can help identify the intricate complex of causes behind peoples' displacement and their choice to migrate in line with 'objective causes': heteropatriarchal, class, and racist violence, forms of discrimination and limited freedoms (Giuliani, 2021; Mezzadra, 2004).

This framework foregrounds the colonial roots of both mobility and place, as well as the fact that adaptive capacity is highly asymmetrical and mediated by one's intersectional position in relation to capital, gender, ethnicity, class, and race (Boas et al., 2019). At the same time, the theory of mobility justice helps us move beyond the residentialist approach to the climate crisis and focus on the 'making of migration' (Tazzioli, 2020) through biopolitical and geopolitical mobility regimes, and the 'apocalyptic perspective' strongly promoted by the media.

Climate-induced mobility is often framed through the lens of displacement, with sudden-onset disasters such as floods or hurricanes and slow-onset processes like desertification or sea-level rise compelling individuals and communities to leave their homes. However, this perspective often neglects the mirror phenomenon of 'trapped populations'—vulnerable groups unable to migrate due to structural barriers such as poverty, lack of resources, or deep socio-cultural ties to place. This dual dynamic of enforced movement and enforced stasis challenges traditional notions of justice, especially when layered across multiple scales—national, international, intergenerational, and cosmological. At the national level, restrictive border policies criminalize climate migrants while also failing to address internal displacement. Internationally, non-binding frameworks like the Global Compact for Migration highlight gaps in accountability for high-emission states that disproportionately contribute to climate-induced displacement. An intergenerational justice lens further complicates this nexus, as current inaction on climate adaptation continues to shift the burden onto future generations. Indigenous cosmologies, which center on reciprocal human-ecosystem relationships, offer crucial alternatives to state-centric models that frequently undermine cultural ties through forced relocations. Addressing climate (im)mobility justice requires an integrated framework that not only ensures safe migration pathways but also supports those unable or unwilling to move while holding states accountable for their role in furthering environmental degradation and displacement.

A second perspective on climate (im)mobilities seen through the lens of climate justice calls attention to those least responsible for the climate crisis but most impacted by its effects. These are the subalterns who cannot speak or whose voices are not heard (Spivak, 1988). In terms of procedural justice, for example—defined as 'justice in the way outcomes are brought about, including who participates and is heard in the processes of decision-making' (Agyeman et al., 2016, p. 2913)—a significant issue of justice for migrants lies in their minimal involvement in the policy decisions that impact their lives (Edge et al., 2020). This concern is especially pertinent for forced migrants and refugees who are often excluded from discussions about fundamental aspects of their displacement and resettlement, including the selection of migration destinations (Atapattu, 2020). The IPCC, for example, has highlighted the situation of refugee camps in Central Africa, which are increasingly affected by rising temperatures (Birkmann et al., 2022). Policies frequently place these camps in vulnerable areas (Owens et al., 2023) and subsequently limit the mobility required of refugees to adapt to life-threatening conditions. These limitations in turn restrict migrants' ability to adapt more broadly, deepening their precarity in host countries where they lack full rights and citizenship. As we have seen in the previous chapter, hospitality is caught between open-ended principles and bounded laws. Derrida (2000) defines it as 'hostipality,' a neologism capturing the legal and ethical tensions between the hostility of restrictive border politics and the gift of unconditional hospitality. At the same time, focusing on the link between climate and mobility in terms of restorative (or corrective) justice—which has emerged as a significant framework in the IPCC AR6 and is in part reflected in climate policy discussions—helps us foreground the principles of fairness, equity, and accountability (Cundill et al., 2021) in global migration governance.[2]

Although we have criticized the neoliberal underpinnings of climate policy negotiations between governments within an international jurisdiction—in which justice issues are often eclipsed—this approach does offer a financial mechanism through which the reception of migrants could be handled through an international Green Climate Fund. As Farber (2008) argues, climate reparations between nations would enable 'creditor countries'—such as small island states—to become more resilient by funding disaster risk reduction, insurance, and adaptation to help people remain in place. Similarly relevant here is the recent proposal to embed climate visa programs within the mechanisms of *Loss and Damage*, which would reflect the emissions of countries when issuing visas to incoming migrants (Cisse et al., 2023).

[2] The IPCC's Sixth Assessment Report (AR6) highlights several exemplary climate policies and strategies, emphasizing the need for both mitigation and adaptation efforts. Available at https://www.ipcc.ch/report/sixth-assessment-report-cycle/

Another pathway to climate compensation is suing oil companies. As Neef and Sheller suggest, under the loss and damage legal approach, 'states could put pressure on companies by adopting legislation making it mandatory for fossil fuel companies to provide compensation through payment of a so-called 'climate damages tax,' […] based upon how much carbon dioxide is embedded within each ton of coal, barrel of oil or cubic liter of gas that each fossil fuel company extracts' (Neef & Sheller, 2024, p. 470).

Ultimately, recognitional justice focuses on addressing the misrecognition or marginalization of certain groups, ensuring they are seen, heard, and valued in society. When extended to the realms of climate and mobility justice, this concept becomes crucial in addressing how marginalized communities experience the impacts of climate change and gain access to just mobility opportunities. Focusing on the intersection of social equity, sustainability, and the need for a more inclusive and responsive society that understands the unique challenges faced by disadvantaged groups, recognitional climate mobility justice offers a framework that addresses forced displacement and international migration, particularly as expected from Small Island Developing States in the coming decades. Displacement from these and other forms of climate-related disasters must always be examined within their broader context. In most cases, an analysis of climate mobility through the lens of racial justice should be linked to other forms of social and economic inequity. For instance, the displacement experienced by Puerto Ricans following Hurricane Maria was significant and affected individuals from a wide range of socioeconomic, racial, and ethnic backgrounds. Nevertheless, those with greater resources were able to evacuate the island before the hurricane's arrival, while others lacked such opportunities (Palinkas, 2020, p.36).

A climate mobility justice lens also recognizes Indigenous people's long-standing mobility practices and capacities and acknowledges the importance of Indigenous self-determination in governing contemporary climate (im)mobilities. This includes the rights of Indigenous people to stay in their ancestral territories and resist attempts of governments and international donors to resettle them against their will and repeat the violence of forced, colonial-era relocations (Yumagulova et al., 2023).

Climate (Im)Mobilities and Narratives

The power of representation is central both to legitimizing contemporary climate (im)mobility regimes and to unveiling the traps and misconceptions laid within the concept of climate (im)mobility justice (Boas et al., 2022). Media narratives of the nexus between climate change and migration (Giacomelli, 2023; Sakellari, 2024; Giacomelli & Cappi, 2025) often rely on alarmist, sensationalist language, and hostile framing. Climate migrants are frequently presented as a threat to global, regional, national, or subnational security—not only by segments of the news media but also by military institutions, researchers, and civil society actors (Bettini, 2013a). Rather than focusing on the social injustice that exacerbates vulnerability to climate change, media in rich and industrialized countries tend to depict cross-border migration as a security implication of climate change (Schäfer et al., 2016). The image of 'waves' of climate migrants threatening the security of countries in the so-called Global North thus feeds and amplifies distorted perceptions, reinforced by the increasing allocation of public funds to border enforcement and securitization rather than to environmental protection.

It is therefore unsurprising that, globally, climate change-induced migration ranks fifth on a list of more than thirty perceived security risks (Bunde et al., 2024). Nor is it surprising that the United States has, for years now, incorporated climate change and environmental migration into its defense and security strategies. As Betsy Hartmann (2010) shows, such dystopian narratives are employed by US defense circles to cast the vulnerable—particularly environmental migrants—as national security threats.

As explored in previous chapters, mainstream discourses on climate change and migration often fail to grasp the complex interdependencies between human mobility patterns, anthropogenic climate change, and the historical and political trajectories of colonial exploitation and border governance mechanisms. This omission flattens the multifaceted nature of these interconnected phenomena as well as their broader socio-political contexts.

Although human mobilities are shaped by varying degrees of power and inequality, when environmental migration enters mainstream media visibility, it is typically framed within a reductive emergency paradigm that oscillates between humanitarian and securitarian logics. The climate-migration nexus is thus either securitized—as a global, or national threat precipitating chaos and conflict (Sakellari, 2024; Warner & Boas, 2017; Peters & Mayhew, 2016)—or humanized as a duty to help vulnerable victims. This paradox between repressive migration governance and humanitarian concerns is what Didier Fassin (2011) terms 'compassionate repression.'

Despite the reality that most climate-related mobility consists of short-distance, internal movements—frequently from rural to urban settings—and when international movements do occur, it is generally regional (Kolmannskog, 2008; Wilkinson et al., 2016)—climate migration has increasingly been framed as a humanitarian issue. It is often construed as the result of a failure in the place of origin and as a burden for the destination. The mainstream narrative that climate change will lead to mass displacement and an unmanageable influx of 'climate refugees' into Europe has been frequently repeated—both in headlines and in the humanitarian sector—in the attempt to impel the Global North to 'do something' about climate change (Randall, 2017). As we will show in Chapter 6 through our analysis of awareness and advocacy campaigns, while humanitarian framing may attract additional funding for the humanitarian sector, it also risks reproducing 'us vs them' binaries and 'savior-victim' logics. In this sense, climate change becomes a vehicle for reinforcing colonial gazes and 'othering processes'—unless reframed through the lens of climate and mobility justice.

In parallel, the securitarian narrative portrays climate change as a threat multiplier of migration—an interpretation that serves specific political interests, including the anti-immigration policy agendas of right-wing and populist governments in different countries. Even though most climate-related migration takes place within country borders, the securitarian narrative of climate migration increasingly justifies militarizing national and regional borders in the global North (Boas et al., 2019).

More broadly, in the early twenty-first century, as the effects of global heating took hold in earnest, the narrative of climate change emerged as a national security strategy. Oil companies like Shell, along with powerful militaries, recognized climate change had the power to destabilize industries, economies, and nations, and started framing climate change as a threat to the status quo. As the Transnational Institute has documented, energy corporations have used their power and wealth to influence policy and create a global energy system that puts their profit-making above all else, even continuing life on earth. These same corporations—responsible for a significant share of the global emissions accelerating climate change—now benefit from securitization discourses that legitimate their expansion. Military planners, weapons manufacturers, and the burgeoning border and surveillance industry are using this narrative to position themselves as providers of 'climate security,' securing increased public funding to make borders more dangerous and inaccessible to migrants. Anti-migrant and white supremacist groups are thrilled at the prospect of world militaries and climate activists adopting their ethnonationalist language and are using this climate 'threat' narrative to stoke fears of uncontrolled immigration and disseminate anti-refugee narratives—with significant rhetorical and electoral success across the Global North.

Although the link between climate and mobility 'needs to be conceptualized as multidimensional and as a continuum across different timescales, spatial scales, and with respect to the level of agency and control that people have over where they live' (Neef et al., 2024, p. 5), these nuances are routinely erased by political, corporate, and media discourses. Not only are some of the actors most vulnerable to climate change categorically unable to migrate—remaining in 'trapped communities'—but whatever mobility they may engage in remains dependent on their social and economic capital (Schewel, 2020; Black & Collyer, 2014). Additionally, the very notion of political agency—whether of those in motion or those who stay—is frequently overlooked (Wiegel et al., 2019)?

Narratives about climate migration inform corresponding mindsets and public responses. Depicting climate migrants as security threats, fuels fear-based reactions and legitimizes securitization and criminalization. The label 'climate refugee,' for example, has been co-opted into security discourses, inciting fears of mass migration and bolstering a 'flooding our shores' rhetoric together with its respective 'fortress' mentalities (Tschakert & Neef, 2022). David Theo Goldberg et al. (2021) identifies in this a deeper anxiety: an out-of-control climate is equated with an out-of-control people, as if both derived from chaotically 'natural' rather than political events. By portraying climate change itself as 'a migration crisis in the making' (Baldwin & Bettini 2017, p. 1), these alarmist narratives are misused to exacerbate fear narratives and strengthen border control in the global North (Durand-Delacre et al. 2021; Lietaer & Durand-Delacre, 2021). Accelerating climate destabilization is closely linked to growing securitization and border enforcement (Delacre, 2023). Not only do the security apparatus in the United States and the EU view climate migration as a major threat and prepare future scenarios in response (Squire

2012), but the resources poured into border enforcement far exceed those allocated to climate mitigation.

The priority is security (von Lucke, 2021) and this can be seen in the increasing militarization of borders (Miller et al., 2021). The EU budget for external border, migration, and asylum management for 2021–2027 has increased 2.6 times to more than 34.9 billion euros, compared to 13 billion euros for 2014–2020. Recent research predicts the annual growth rate of the global border security market to be between 7.2 and 8.6 percent ($65–68 million) in 2025.[3] Oxfam's research on funding allocated through the *EU Trust Fund for Africa* reveals that ODA spending is increasingly linked to the EU's desire to stop irregular migration (Raty & Shilhav, 2020). Copious research (Nshimbi & Moyo, 2016; Mager, 2018; Zardo, 2022) and investigations, such as ActionAid's *The Big Wall*, have tried to unmask the public resources mobilized in support of border externalization policies and the construction of the 'big wall' to defend 'Fortress Europe.' Although partial estimates on specific sectors and countries do exist, no official data is currently available for the purposes of collecting, organizing, and making information on the distribution and total amount of these resources accessible.

A recent report from the Transnational Institute (Miller et al., 2021) finds that the world's largest emitter states spend, on average, 2.3 times more on border militarization than on climate mitigation. Seven countries responsible for 48% of world-historical emissions, collectively spent at least twice as much on border protection (more than $33.1 billion) between 2013 and 2018 than on climate change mitigation ($14.4 billion). According to the Climate Equity Calculator, the United States currently accounts for 30% of global emissions, followed by the EU (18%), China (16%), Japan (4%), Canada (3%), and Australia (2%).

As Nick Buxton of the Transnational Institute notes: 'Money is being spent on militarization rather than on climate action, though the climate crisis is the biggest [common] security threat that we face today.'[4] Indeed, wealthy nations—especially NATO members—including the UK, Canada, and the US—the world's most militarized nation and largest institutional emitter—accounted for more than half of all global military

[3] https://www.marketresearchfuture.com/.

[4] https://www.theguardian.com/environment/2023/dec/02/cop28-climate-change-military-funds

spending in 2022, with current budgets expected to rise further. In 2023, NATO members made an enduring commitment to invest at least 2% of their national budgets in their militaries. If that promise is fulfilled, the researchers calculated that, based on United Nations data, by 2028 the additional military funding would be enough to pay for seven years' worth of climate adaptation in all the hardest-hit countries combined. Research shows that increased militarization harms climate-affected countries in other ways, too. For instance, they found that NATO members are currently selling arms to 39 of the 40 most climate-vulnerable countries; 17 of which are already in armed conflict.

Underscoring that the world's militaries produce at least 5.5% of greenhouse gas emissions—more than the total footprint of Japan—the Transnational Institute calls for the de-militarisation and de-securitisation of a whole host of issues, including climate change.

At Cop28 in Dubai, several climate justice organizations demanded that wealthy nations redirect at least 5% of their military budgets toward climate finance. 'It's a real perversion,' Buxton argued, 'rather than providing support for communities we are actually making the situation more dangerous and more unstable.' To make the world safer, we need a new narrative of collective security—one based on peace, well-being, care, solidarity, human rights, and democracy, rather than on militarization.

As Mark Neocleous (2019, p. 31) argues, with its management of terror, 'the one thing the modern state can and does guarantee is the perpetual "insecurity" of the capitalist order.' In this sense, climate migration is one of the many manufactured insecurities tailored to the needs of capitalism and its elites. Its treatment as a problem to be solved (Bettini, 2019), leads to both its pathologization and depoliticization' (Baldwin & Bettini, 2017). The view of environmental migration and human mobility as pathology ultimately has its roots in the modern Western (and colonial) biopolitical nexus of mobility-territory-sovereignty, which presupposes an organic link between nation-population and state-territory (Agamben, 1998). Depoliticization, on the other hand, reduces the narrative to the *hic-et-nunc* of the present, without bothering to analyze the root causes of the two phenomena and the nexus between them. Thus, rather than recognizing the ecological crisis as a crisis of justice, the capitalist system rooted in nation-states enshrines security as the control of privilege through coercion. This security, however, is grounded in desperation and fear, rather than in compassion and concern. In doing so, the preservation of an unequal and ecocidal status quo takes precedence over any commitment to a just future based on solidarity and redistribution.

The Racialized Figure of the Climate Migrant

The racialized figure of the climate migrant emerges as a paradoxical construct within media, policy, and academic discourses: simultaneously hypervisible yet unverifiable—an entity that 'exists, but its existence can never be objectively verified' (Baldwin, 2013, p. 1484). Despite climate change being a global phenomenon, environmental migrants are predominantly depicted as racialized subjects from the Global South whose movement toward the Global North is framed through securitarian and alarmist narratives of crisis and invasion (Baldwin et al., 2014; Buxton et al., 2016; Telford, 2018). This racialization is underpinned by three interrelated elements: naturalization, loss of political status, and ambiguity (Baldwin, 2013). Naturalization dehistoricizes both climate change and migration, detaching environmental migrants from political and historical contexts, thereby rendering them either passive victims in need of rescue or threats to Western stability (Musarò, 2017). The loss of political status further exacerbates their marginalization, positioning them as stateless non-citizens, devoid of rights and protection under international law, reinforcing xenophobic and exclusionary politics (Agamben, 1998). Ambiguity—fueled by alarmist framings and imprecise definitions—contributes to the dehumanization of climate migrants, transforming them into abstract figures that symbolize disorder and the dissolution of borders, particularly within the Western imaginary (Baldwin, 2013; Biermann & Boas, 2010; Gill, 2010; Turton, 2003). This process is deeply entangled with media representations that capitalize on fear and crisis to shape public perceptions and policy responses (Russo & Wodak, 2019). Positioned as a liminal figure, the climate migrant embodies anxieties about control and the destabilization of borders, reflecting what Baldwin (2013, 2022) describes as 'white anxieties' linked to impending disorder. Moreover, the framing of climate migration as a security threat often suppresses the agency and self-representation of those most affected, reinforcing their marginalization (Dreher & Voyer, 2015; Jaworski, 1993).

As Chouliaraki (2006) argues, the dehumanization of suffering individuals erases their political agency, reducing them to passive subjects of humanitarian discourse rather than active participants in their own future-making. Ultimately, the racialized figure of the climate migrant reflects broader Western fantasies of border control and containment (Bettini, 2013b), thus calling for a fundamental reimagining of climate mobility beyond securitized and racialized framings—one that embraces alternative narratives of agency, resilience, and justice.

So much for transnational movers, but what of internal climate migrants in this story? The discourse on climate-induced migration reveals a stark asymmetry in how internal displacement is recognized and studied across different world regions, thus reinforcing the racialized figure of the climate migrant as a mobility monolith (Durand-Delacre, 2023). Reports such as the World Bank's Groundswell focus on internal displacement in regions like Sub-Saharan Africa, South Asia, and Latin America, while excluding from their analysis high-income countries like the US, Canada, the nations of Western Europe, Australia, and New Zealand.[5] This exclusion is treated as self-evident, with no explicit justification provided, unlike the detailed reasoning offered for omitting small island developing states (SIDS) or countries in the Middle East. Such methodological choices obscure the linkages between internal and international migration, despite scholarly calls to explore how these dynamics might complement or substitute one another. Moreover, a systematic review of the CliMig database[6] highlights that most case studies focus on the so-called Global South, even though the majority of researchers are affiliated with institutions in high-income countries. While the field has moved away from overtly securitized framings of migration as a 'threat,' it continues to reflect an implicit assumption that climate displacement is primarily a concern of populations in the Global South.

This imbalance not only marginalizes the experiences of those affected by climate change in high-income countries but also re-racializes the

[5] https://researchinginternaldisplacement.org/short_pieces/climate-related-mobility-mostly-occurs-within-borders-disentangling-factual-truth-from-a-convenient-narrative/.

[6] All information are available at CliMoHub webpage: https://www.climohub.org/. CliMoHub is a Knowledge Hubon Environmental and Climate (Im)Mobilities,based and operated at the University of Vienna. The CliMoHub provides access to a wide range of qualitative and quantitative secondary data (sets), literature, case studies and visualisations with the aim of fostering a deeper understanding of the complex and interconnected issues at the intersection of climate change and human (im)mobility—including migration.

figure of the climate migrant as predominantly belonging to poorer, non-Western regions—thus reinforcing uneven power dynamics in knowledge production about climate mobility.

Other Narratives Arising

As we have seen, the issue of climate (im)mobilities is accompanied by a range of biases, misunderstandings, and, above all, toxic and distorted narratives that legitimize unjust policies. Changing the narratives can generate new conversations—and potentially new policies—around the responsible reception of climate migrants, and what it might mean to envision open borders or a more hospitable environment in a world increasingly shaped by climate extremes. But how can we raise awareness of climate (im)mobilities without reproducing harmful narratives? What elements are necessary to build more accurate, equitable representations? How can we challenge dominant discourses through justice-oriented approaches that foster solidarity, uphold human rights, and advance intersectional climate initiatives?

As climate change alters global mobility trends, it is essential to prioritize equity and human rights in the policy strategies and the stories we tell about climate (im)mobilities. First, this requires cross-sectoral collaboration—among social movements, media, academia, labor unions, and numerous other organizations in civil society. Many initiatives already exist, from the Environmental and Climate Mobility Network,[7] a transdisciplinary network connecting people working on migration and human im/mobility in the context of environmental and climate change, to the Environmental Justice Foundation (EJF), which focuses on the intersection of human rights and environmental protection while advocating for the rights of climate refugee and displaced populations. Likewise Refugees International provides policy recommendations to address the needs of climate migrants. The Nansen Initiative has similarly advocated for legal frameworks that protect those dislocated by environmental change. All these organizations are influential in promoting understanding, raising awareness, and driving international action to address the complex issue of climate (im)mobility.

[7] https://www.climatemobilities.network/

Among the tools produced to reframe public understanding, is the brief 'Connecting Climate Justice & Migrant Justice: A Guide to Countering Dangerous Narratives',[8] launched by the platform Migrants Organise[9] in collaboration with other advocacy groups.[10]

As these and many other activist organizations underscore, in guiding the conversation toward a more humane and compassionate perspective, it is essential to advocate for values that tease out the intrinsic connection between the individuals impacted by climate change and, in case of displacement, the communities that host them. By emphasizing shared values and acknowledging the global implications of climate change, we can cultivate a narrative that resonates with different audiences on a deeper, more empathetic level. Stories should be grounded in distinct imaginaries—anchored in local experiences, responsive to material realities, and oriented toward community-based solutions.

Equally important is naming those who profit from the perpetuation of harmful migration systems. This includes corporations that benefit from border control and surveillance industries, as well as fossil fuel companies that hinder the transition to sustainable energy solutions. By identifying these 'villains,' we can redirect the narrative toward actual, genuine threats and better advocate for necessary policy changes. Climate (im)mobility narratives are best served by decolonial and intersectional approaches that unmask(in)justice and naming the culpable.

As the Artists Circle suggests, when telling climate migration stories, we must ask: 'Who benefits or loses from telling the story? Who has the right to a story? What are the dynamics of power in the distribution of a story? How does the story's form change how we understand it?'.[11]

Following the justice-oriented approaches outlined earlier in this chapter, we must also ask: Why do we tell the story of the individual forced to leave and not the story of the oil company or lobbyists who

[8] https://www.migrantsorganise.org/connecting-climate-justice-migrant-justice-a-guide-to-countering-dangerous-narratives-report-launch/

[9] https://www.migrantsorganise.org/about-us/

[10] Some of the associations include: 350.org, the Climate and Migration Coalition, the Climate Justice Coalition, the Climate Justice Collaborative NPNA, Common Defense, Friends of the Earth US, International Refugee Assistance Project (IRAP), the Joint Council for the Welfare of Immigrants, and Pacific Island Students Fighting Climate Change.

[11] https://belonging.berkeley.edu/artist-circle/why-artists.html

created the conditions for that displacement? Why are we not linking the stories of Black, Brown, and Indigenous people to a deeper moral case for climate reparations based on the long-term effects of systems of colonialism, and racial capitalism? As Ranganathan and Bratman (2021) argue, the call for 'abolitionist climate justice' should embed climate debt in longer historical time frames of infrastructural repair, land return, and social reparations.

Under current legal frameworks and migration/refugee policies, the debate is mainly focused on international protection in the context of adverse effects of climate change or disasters. The international refugee regime operates within a world of sovereign states, with international refugee instruments that preserve the right of a nation-state to determine who may traverse and reside within its borders. How can we challenge sedentary epistemologies and explore more cosmopolitan narratives? What solutions would arise from applying a different storytelling ethic in media, institutional communication, or in art?

Almost a decade ago, Ransal-Cooper et al. argued that a variety of actors had interpreted and responded to the issue of climate (im)mobility[12] through four distinct frames: victims, security threats, adaptive agents, and political subjects. Although not always consistently deployed by the same actors, and often hybridized, the authors contended: 'the four framings highlight different political, ideological and practical beliefs associated with particular interests and normative assumptions' (2015, p. 106).

Can we still recognize these four framing typologies today? Do they still hold as cognitive filters for sense-making? How might we update and enrich their application within current popular and academic discourse?

Moving on from the four distinct frames identified by Ransan-Cooper et al. (2015), contemporary perspectives should call for the incorporation of justice-oriented, decolonial, intersectional, and feminist approaches to climate (im)mobility. These expanded framings for 'coping with crisis' emerge through key elements of narratives, their geographical contexts, temporal dimensions, and the roles of the protagonists involved. Victimization status should extend beyond human subjects, incorporating more-than-human entities in accordance with posthumanism's

[12] In their article they define the issue as 'environmental migration'.

challenge to anthropocentric narratives (Clark et al., 2023). From a feminist perspective, the notion of vulnerability in a world marked by climate change should not be placed in opposition to people's collective agency (Sahraoui, 2023).

Moreover, securitization discourse, traditionally used to paint migrants as threats (Fassin & Pandolfi, 2010), should be reframed to highlight structural injustices and the colonial legacies of border regimes (Mbembe, 2019) while emphasizing that security can either operate as a mechanism of repression and control aimed at maintaining the privilege of a few or as a form of solidarity and care for all living beings, human and non-human. In the first case, security would resonate with injustice and be attached to negative emotions such as fear and despair. In the second iteration, security echoes calls for collective care and can be conveyed through a sense of shared vulnerability and hope.

The adaptive agent has also evolved, recognizing that migration is not merely reactive, but can be a proactive strategy bolstered by governance support—although, to be sure, state policies remain largely inadequate (Scott, 2023). Lastly, political subjectivity must be expanded to include climate displacement in wealthier nations, challenging the default assumption that climate migration is always a problem of the Global South.

Together, these enriched frames move us beyond securitarian, crisis-driven paradigms and toward narrative futures grounded in agency-affirming tellings to interconnected vulnerabilities, and intersectional and decolonial depictions of climate (im)mobility.

Conclusions

As we conclude this chapter, it becomes increasingly evident that the intricate relationship between climate change and human mobility is a complex tapestry woven from threads of socioeconomic, political, and geographical factors. For centuries, environmental changes have shaped human mobility, with seasonal movements integral to societal structures (Ionesco et al., 2017). However, it is only in recent decades that the international community has explicitly acknowledged the deep interconnection between climate change and human mobility, reflecting the urgency of the issue (Laczko & Aghazarm, 2009; Piguet, 2022).

Climate (im)mobility is often portrayed as a mass movement from the Global South to the Global North—a script generating securitized and racialized fears. Yet this narrative obscures the reality that most environmental migrations occur within national borders or to neighboring countries (Piguet, 2013). The International Organization for Migration (IOM) defines climate migration as the movement of individuals due to environmental changes, yet the term 'climate refugee' remains problematic due to the lack of international legal protection (IOM, 2019; IPCC, 2014). This legal vacuum leaves individuals vulnerable, often classified as internally displaced persons rather than refugees, highlighting the need for a more inclusive and justice-oriented approach to addressing climate-induced mobilities.

The prevailing narratives surrounding climate migration frequently reinforce a dichotomy between security and justice. In many cases, climate-induced (im)mobilities are treated as crises to be contained rather than as phenomena to be addressed through equitable policies. By unpacking procedural, restorative, and recognitional justice frameworks, we reveal how climate-induced (im)mobilities are embedded in both historical and contemporary power structures. The increasing militarization of borders and disproportionate spending on security over climate (im)mobilities illustrate the need for a profound paradigm shift. Rather than fortifying borders against climate migrants, global actors must commit to policies that uphold mobility as a right and recognize the obligations of historically high-emitting countries toward those most affected by climate change.

This chapter has demonstrated the importance of moving beyond securitized framings and adopting an intersectional approach to climate (im)mobilities justice. Dominant narratives often perpetuate stereotypes and frame climate change as a national security threat, which in turn has led to increased border enforcement and militarization, diverting resources away from climate mitigation efforts (Baldwin & Bettini, 2017; Durand-Delacre et al., 2021). This trend is exemplified by the disproportionate allocation of funds to border security compared to climate change mitigation, with major emitters spending significantly more on border protection than on addressing the root causes of climate displacement (Transnational Institute, 2021).

To move beyond these securitized narratives, it is essential to adopt a multidimensional approach to climate mobility justice. This involves recognizing the agency and self-determination of communities affected by climate change, as well as advocating for policies that address the root causes of displacement (Farbotko, 2022a). By embracing alternative theoretical frameworks such as mobility justice and the coloniality of power, we can begin to unravel the systemic injustices and inequalities that underpin climate-induced mobilities (Quijano, 2000; Sheller, 2020).

Climate-induced (im)mobilities, we argued, are not merely outcomes of environmental change but are deeply intertwined with capitalist-driven resource extraction, land dispossession, and exploitative labor markets that push marginalized communities into precarious conditions (Malm, 2016). Simultaneously, restrictive border regimes and securitized migration policies criminalize the movement of the displaced, reinscribing global hierarchies of mobility and immobility (Boas et al., 2019).

Ultimately, addressing climate mobility as a justice issue requires a profound shift in our collective understanding. The scalar nature of justice in climate (im)mobilities challenges us to think beyond traditional boundaries, to envision a world where the fates of distant communities are inextricably linked, where the actions of today reverberate through generations, and where our sense of belonging extends beyond national borders toward a truly cosmopolitan ethic. Our examination of competing security narratives has exposed the stark contrast between responses based on militarization and border control, and those grounded in care, solidarity, and cosmopolitan justice.

We have also emphasized the critical role of subjective experiences, emotions, and imaginaries in shaping perceptions and responses to climate (im)mobilities. Recognizing the diversity of perspectives and the power of narrative is essential for developing more inclusive and effective ways of addressing this global challenge. It demands that we prioritize human rights, equity, and solidarity over securitized and militarized approaches (Klein, 2015). By fostering a narrative that resonates with empathy and inclusivity, we can work toward more transformative solutions that acknowledge the intrinsic connection between individuals impacted by climate change and the communities that host them (Farbotko et al., 2022b).

To illuminate how a diverse range of stakeholders delineate the parameters of climate (im)mobilities, thereby shaping interpretations and organizing perceptions and imaginaries more broadly, in the next chapters we will explore how international nongovernmental (INGOs) and governmental organizations (such as IOM, UNHCR), as well as contemporary artists, are symbolically giving a human face to climate mobility regimes.

We will focus on these actors and their drive toward justice for several reasons.

Alongside journalists and human rights professionals, INGOs, the UN, and other organizations play a crucial role in shaping the narratives around the issue by informing or influencing behaviors in large audiences, amplifying the voices of vulnerable communities, conducting research, and advocating for policy change. They help raise awareness of the links between environmental degradation and migration, ensuring that these issues are included in global discussions. They can tell the truth about the dynamics of power that inflict indignation and humiliation while intensifying twenty-first-century risks. They can play a pivotal role in cultivating constructive change in human engagement by widening our lens on climate mobility regimes. INGOs, the UN, and other organizations can also offer practical solutions and support to displaced populations, highlight the human rights aspects of climate (im)mobilities, and push for sustainable and just responses from governments and international bodies. Through their climate adaptation initiatives, they also help reimagine what security and protection might mean in an era of climate volatility. By funding climate adaptation and advocating for safe, dignified migration pathways, these grassroots organizations challenge dominant scripts and help push for urgent action.

As we will show in the following chapter, through their awareness campaigns, these actors contribute to the constant processes of construction (bordering), deconstruction (debordering), and reconstruction (rebordering) of the border itself (Brambilla, 2015). Within this framework, where border construction, representation, and policing converge in the concept of *borderscape* (Agier, 2016), we intend to examine how recent awareness campaigns operate as symbolic interventions aimed at regulating the mobility of certain people.

Finally, the book's concluding chapter will turn to contemporary artistic practices offering a new perspective on the dynamics of climate and mobility justice. Our aim is to widen the claim of justice by pushing the imagination beyond geopolitical realism and the restrictive legal frameworks of the present. The arts are powerful tools for raising awareness, evoking emotional responses, and inspiring action. Through various mediums—such as visual art, music, literature, theatre, and performance—artists convey the urgent realities of climate and mobility justice, making abstract scientific data more relatable and personal. Art can help humanize the impacts and complexities of climate (im)mobility regimes, giving them a face and a story, which often resonates more deeply than statistics alone. Artists can challenge dominant narratives and open up alternative spaces of imagination, representation, and action (Mazzara, 2019; Musarò et al., 2022).

Art is also a crucial outlet for activism: sparking conversations, it challenges perceptions and fosters a collective desire for meaningful change. It is not surprising that activists for climate change are increasingly turning to art and museums to subvert traditional structures of power. As powerful platforms for protest, art and museums offer good opportunities to engage different audiences emotionally, creatively, and intellectually. We contend that when artists express their fullest sense of belonging, it is not confined to their nation-states, nor to this planet, 'as their exploration of a connection to the commons extends all the way to the cosmos' (Papastergiadis, 2023, p. 156). Art does not merely reflect the world—it also makes new ones (Goodman, 1978), expanding our horizon of possibility and cultivating the imaginative groundwork for more just futures.

References

Agamben, G. (1998). *Sovereign power and bare life*. Stanford University Press.

Agyeman, J., Schlosberg, D., Craven, L., & Matthews, C. (2016). Trends and directions in environmental justice: From inequity to everyday life, community, and just Sustainabilities. *Annual Review of Environment and Resources, 41*(1), 321–340. https://doi.org/10.1146/annurev-environ-110615-090052

Ahsan, R. (2019). Climate-induced migration: Impacts on social structures and justice in Bangladesh. *South Asia Research, 39*(2), 184–201. https://doi.org/10.1177/0262728019842968

Aidoo, F. S. (2021). Architectures of mis/managed retreat: Black land loss to green housing gains. *Journal of Environmental Studies and Sciences, 11*(3), 451–464. https://doi.org/10.1007/s13412-021-00684-3

Alaimo, S. (2010). Bodily natures: Science, environment, and the material self. Indiana university press.

Armiero, M. (2021). *L'era degli scarti: Cronache dal Wasteocene, la discarica globale*. Einaudi.

Arnall, A., Hilson, C., & McKinnon, C. (2019). Climate displacement and resettlement: The importance of claims-making 'from below.' Climate Policy, 19(6), 665–671. doi:https://doi.org/10.1080/14693062.2019.1570065.

Atapattu, S. (2020). Climate change and displacement: Protecting 'climate refugees' within a framework of justice and human rights. *Journal of Human Rights and the Environment, 11*(1), 86–113. https://doi.org/10.4337/jhre.2020.01.04

Baldwin, A. (2013). Racialisation and the figure of the climate-change migrant. *Environment and Planning A: Economy and Space, 45*(6), 1474–1490. https://doi.org/10.1068/a45388

Baldwin, A. (2022). *The other of climate change: Racial futurism, migration, humanism*. Rowman & Littlefield.

Baldwin, A., & Bettini, G. (Eds.). (2017). *Life adrift: Climate change, migration, critique*. Rowman & Littlefield International.

Baldwin, A., Fröhlich, C., & Rothe, D. (2019). From climate migration to anthropocene mobilities: Shifting the debate. *Mobilities, 14*(3), 289–297. https://doi.org/10.1080/17450101.2019.1620510

Baldwin, A., Methmann, C., & Rothe, D. (2014). Securitizing 'climate refugees': The futurology of climate-induced migration. *Critical Studies on Security, 2*(2), 121–130. https://doi.org/10.1080/21624887.2014.943570

Barca, S. (2020). *Forces of reproduction: Notes for a counter-hegemonic anthropocene*. Cambridge University Press.

Bayrak, M. M., Marks, D., & Hauser, L. T. (2022). Disentangling the concepts of global climate change, adaptation, and human mobility: A political-ecological exploration in Vietnam's Mekong Delta. *Climate and Development, 14*(10), 935–944. https://doi.org/10.1080/17565529.2022.2028596

Bettini, G. (2013a). Climate Barbarians at the Gate? A critique of apocalyptic narratives on 'climate refugees.' Geoforum, 45, 63–72. doi:https://doi.org/10.1016/j.geoforum.2012.09.009.

Bettini, G. (2013b). (in)convenient convergences: "climate refugees", apocalyptic discourses and the depoliticization of the debate on climate-induced migration. In C. Methmann, D. Rothe, & B. Stephan (Eds.), *Interpretive approaches to global climate governance* (pp. 122–136). Routledge.

Bettini, G. (2019). And yet it moves! (climate) migration as a symptom in the Anthropocene. *Mobilities, 14*(3), 336–350. https://doi.org/10.1080/17450101.2019.1612613

Biermann, F., & Boas, I. (2010). Preparing for a warmer world: Towards a global governance system to protect climate refugees. *Global Environmental Politics, 10*(1), 60–88. https://doi.org/10.1162/glep.2010.10.1.60

Birkmann, J., Jamshed, A., McMillan, J. M., Feldmeyer, D., Totin, E., Solecki, W., Ibrahim, Z. Z., Roberts, D., Kerr, R. B., Poertner, H.-O., Pelling, M., Djalante, R., Garschagen, M., Leal Filho, W., Guha-Sapir, D., & Alegría, A. (2022). Understanding human vulnerability to climate change: A global perspective on index validation for adaptation planning. *Science of the*

Total Environment, *803*, 150065. https://doi.org/10.1016/j.scitotenv.2021.150065

Black, R., & Collyer, M. (2014). Populations "trapped" at times of crisis. *Forced Migration Review*, *45*, 52–56.

Bleich, E., Bloemraad, I., & De Graauw, E. (2015). Migrants, minorities and the media: Information, representations and participation in the public sphere. *Journal of Ethnic and Migration Studies*, *41*(6), 857–873. https://doi.org/10.1080/1369183X.2014.1002197

Boas, I., Farbotko, C., Adams, H., Sterly, H., Bush, S., van der Geest, K., Wiegel, H., Ashraf, H., Baldwin, A., Bettini, G., Blondin, S., de Bruijn, M., Durand-Delacre, D., Fröhlich, C., Gioli, G., Guaita, L., Hut, E., Jarawura, F. X., Lamers, M., et al. (2019). Climate migration myths. *Nature Climate Change*, *9*(12), 901–903. https://doi.org/10.1038/s41558-019-0633-3

Boas, I., Kloppenburg, S., Van Leeuwen, J., & Lamers, M. (2018). Environmental Mobilities: An alternative lens to global environmental governance. *Global Environmental Politics*, *18*(4), 107–126. https://doi.org/10.1162/glep_a_00482

Boas, I., Wiegel, H., Farbotko, C., Warner, J., & Sheller, M. (2022). Climate mobilities: Migration, im/mobilities and mobility regimes in a changing climate. *Journal of Ethnic and Migration Studies*, *48*(14), 3365–3379. https://doi.org/10.1080/1369183X.2022.2066264

Brambilla, C. (2015). Exploring the critical potential of the Borderscapes concept. *Geopolitics*, *20*(1), 14–34. https://doi.org/10.1080/14650045.2014.884561

Bunde, T., Eisentraut, S., & Schütte, L. (2024). Munich security report 2024: Lose-lose? *Munich Security Conference. https://.* https://doi.org/10.47342/BMQK9457

Buxton, N., Hayes, B., & George, S. (Eds.). (2016). *The secure and the dispossessed: How the military and corporations are shaping a climate-changed world.* Pluto Press.

Chouliaraki, L. (2006). Towards an analytics of mediation. *Critical Discourse Studies*, *3*(2), 153–178. https://doi.org/10.1080/17405900600908095

Cisse, G., & et al. (2023). Health, wellbeing, and the changing structure of communities. In H. O. Pörtner, D. C. Roberts, M. Tignor, E. S. Poloczanska, K. Mintenbeck, A. Alegrìa, M. Craig, S. Langsdorf, S. Löschke, V. Möller, A. Okem, & B. Rama (Eds.), *Climate Change 2022—Impacts, Adaptation and Vulnerability: Working Group II Contribution to the Sixth Assessment Report of the Intergovernmental Panel on Climate Change* (1st ed., pp. 1041–1170). Cambridge University Press. doi:https://doi.org/10.1017/9781009325844.

Clark, P. W., D'Amato, A. W., Palik, B. J., Woodall, C. W., Dubuque, P. A., Edge, G. J., Hartman, J. P., Fitts, L. A., Janowiak, M. K., Harris, L. B., Montgomery, R. A., Reinikainen, M. R., & Zimmerman, C. L. (2023). A

lack of ecological diversity in forest nurseries limits the achievement of tree-planting objectives in response to global change. *Bio Science, 73*(8), 575–586. https://doi.org/10.1093/biosci/biad049
Clement, V., Rigaud, K. K., de Sherbinin, A., Jones, B., Adamo, S., Schewe, J., Sadiq, N., & Shabahat, E. (2021). Groundswell part 2: Acting on internal climate migration. *World Bank*. https://openknowledge.worldbank.org/handle/10986/36248
Cresswell, T. (2010). Towards a politics of mobility. *Environment and Planning D: Society and Space, 28*(1), 17–31. https://doi.org/10.1068/d11407
Cundill, G., Singh, C., Adger, W. N., Safra De Campos, R., Vincent, K., Tebboth, M., & Maharjan, A. (2021). Toward a climate mobilities research agenda: Intersectionality, immobility, and policy responses. *Global Environmental Change, 69*, 102315. https://doi.org/10.1016/j.gloenvcha.2021.102315
De Sherbinin, A., Grace, K., McDermid, S., Van Der Geest, K., Puma, M. J., & Bell, A. (2022). Migration theory in climate mobility research. *Frontiers in Climate, 4*, 882343. https://doi.org/10.3389/fclim.2022.882343
Demos, T. J. (2016). *Decolonizing nature: Contemporary art and the politics of ecology*. Sternberg Press.
Derrida, J. (2000). *HOSTIPITALITY. Angelaki, 5*(3), 3–18. https://doi.org/10.1080/09697250020034706
Dreher, T., & Voyer, M. (2015). Climate refugees or migrants? Contesting media frames on climate justice in the Pacific. *Environmental Communication, 9*(1), 58–76. https://doi.org/10.1080/17524032.2014.932818
Durand-Delacre, D., Bettini, G., Nash, S. L., Sterly, H., Gioli, G., Hut, E., Boas, I., Farbotko, C., Sakdapolrak, P., de Bruijn, M., Tripathy Furlong, B., van der Geest, K., Lietaer, S., & Hulme, M. (2021). 6. Climate migration is about people, not numbers. In S. Böhm & S. Sullivan (Eds.), *Negotiating climate change in crisis* (pp. 63–82). Open Book Publishers. https://doi.org/10.11647/obp.0265.06
Duvat, V. K. E., Magnan, A. K., Perry, C. T., Spencer, T., Bell, J. D., Wabnitz, C. C. C., Webb, A. P., White, I., McInnes, K. L., Gattuso, J., Graham, N. A. J., Nunn, P. D., & Le Cozannet, G. (2021). Risks to future atoll habitability from climate-driven environmental changes. *WIREs Climate Change, 12*(3), e700. https://doi.org/10.1002/wcc.700
Edge, S., Brown, E. L., Ghosh, S., & Murnaghan, A. M. (2020). Procedural environmental (in)justice at multiple scales: Examining immigrant advocacy for improved living conditions. *Local Environment, 25*(9), 666–680. https://doi.org/10.1080/13549839.2020.1812554
Everuss, L. (2023). Using mobilities theory to study the nexus between climate change and human movement. *Australian Geographer, 54*(4), 449–458. https://doi.org/10.1080/00049182.2023.2205999

Farber, D. (2008). The case for climate compensation: Justice for climate change victims in a complex world. *Utah Law Review, 2*, 377–413.

Farbotko, C. (2022). Anti-displacement mobilities and re-emplacements: Alternative climate mobilities in Funafala. *Journal of Ethnic and Migration Studies, 48*(14), 3380–3396. https://doi.org/10.1080/1369183X.2022.2066259

Farbotko, C., Kitara, T., Dun, O., & Evans, C. (2022a). A climate justice perspective on international labour migration and climate change adaptation among Tuvaluan workers. *Oxford open. Climate Change, 2*(1), kgac 002. https://doi.org/10.1093/oxfclm/kgac002

Farbotko, C., Thornton, F., Mayrhofer, M., & Hermann, E. (2022b). Climate mobilities, rights and justice: Complexities and particularities. *Frontiers in Climate, 4*, 1026486. https://doi.org/10.3389/fclim.2022.1026486

Fassin, D. (2011). Humanitarian reason: A moral history of the present. *University of California Press*. https://doi.org/10.1525/california/9780520271166.001.0001

Fassin, D., & Pandolfi, M. (Eds.). (2010). *Contemporary states of emergency: The politics of military and humanitarian interventions*. Zone Books.

Flavell, A., Milan, A., & Melde, S. (2020). *Migration, environment and climate change: Literature review* (first report in the migration, Environment and Climate Change Series). Umweltbundesamt.

Giacomelli, E. (2023). *Panicocene: Narrazioni su cambiamenti climatici, regimi di mobilità e migrazioni ambientali*. Franco Angeli.

Giacomelli, E., & Cappi, V. (2025). 'Othering' in the Panicocene: Intergovernmental and non-governmental awareness campaigns on climate change induced-migration. *Journal of Ethnic and Migration Studies, 1–19*. https://doi.org/10.1080/1369183X.2025.2468287

Giacomelli, E., Magnani, E., Musarò, P., & Walker, S. (2023). Crisi climatica, mobilità e giustizia sociale: Voci e storie dal Senegal. *DeriveApprodi*.

Gill, N. (2010). 'Environmental refugees': Key debates and the contributions of geographers: Environmental refugees. *Geography Compass, 4*(7), 861–871. https://doi.org/10.1111/j.1749-8198.2010.00336.x

Giuliani, G. (2021). *Monsters, catastrophes and the anthropocene: A postcolonial critique*. Routledge.

Goldberg, M. H., Gustafson, A., Rosenthal, S. A., & Leiserowitz, A. (2021). Shifting republican views on climate change through targeted advertising. *Nature Climate Change, 11*(7), 573–577. https://doi.org/10.1038/s41558-021-01070-1

Gonzalez, C. G. (2021). Racial capitalism, climate justice, and climate displacement. *Oñati Socio-Legal Series, 11*(1), 108–147. https://doi.org/10.35295/osls.iisl/0000-0000-0000-1137

Goodman, N. (1978). *Ways of worldmaking*. Hackett Pub.

Hamilton, C., Bonneuil, C., & Gemenne, F. (Eds.). (2015). *The Anthropocene and the global environmental crisis: Rethinking modernity in a new epoch* (1st ed.). Routledge. https://doi.org/10.4324/9781315743424

Haraway, D. (2015). Anthropocene, Capitalocene, Plantationocene, Chthulucene: Making kin. *Environmental Humanities, 6*(1), 159–165. https://doi.org/10.1215/22011919-3615934

Hartmann, B. (2010). Rethinking climate refugees and climate conflict: Rhetoric, reality and the politics of policy discourse. *Journal of International Development, 22*(2), 233–246. https://doi.org/10.1002/jid.1676

Hinnawi, E. (with United Nations Environment Programme). (1985). Environmental refugees. United Nations Environment Programme.

Horton, R. M., De Sherbinin, A., Wrathall, D., & Oppenheimer, M. (2021). Assessing human habitability and migration. *Science, 372*(6548), 1279–1283. https://doi.org/10.1126/science.abi8603

Intergovernmental Panel on Climate Change (IPCC). (2014). *IPCC Climate Change 2014: Synthesis Report* (CoreWriting Team; Pachauri, R.K.; Meyer, L.A. (Eds.) Contribution of Working Groups I, II and III to the Fifth Assessment Report of the Intergovernmental Panel on Climate Change). IPCC.

IOM. (2019). IOM global report 2019 OPERATIONS and EMERGENCIES. https://www.iom.int/sites/g/files/tmzbdl486/files/documents/IOM-Global-Report-2019-Operations-and-Emergencies.pdf

Ionesco, D., Mokhnacheva, D., & Gemenne, F. (2017). *The atlas of environmental migration (first published 2017) [map]*. Routledge.

Intergovernmental Panel on Climate Change (IPCC). (1990). *Climate Change: The IPCC 1990 and 1992 Assessments*. https://www.ipcc.ch/report/climate-change-the-ipcc-1990-and-1992-assessments/

Jaworski, A. (1993). *The power of silence: Social and pragmatic perspectives*. Sage.

Jessee, N. (2022). Reshaping Louisiana's coastal frontier: Managed retreat as colonial decontextualization. *Journal of political. Ecology, 29*(1). https://doi.org/10.2458/jpe.2835

Kelman, I. (2020). *Disaster by choice: How our actions turn natural hazards into catastrophes* (First ed.). Oxford University Press.

Klein, N. (2015). *This changes everything: Capitalism vs. the climate*. Penguin books.

Kolmannskog, V. K. K. (2008). *Future floods of refugees: A comment on climate change, conflict and forced migration*. Norwegian Refugee Council.

Laczko, F., & Aghazarm, C. (Eds.). (2009). Migration, environment and climate change: Assessing the evidence. Internat. Organization for Migration.

Lietaer, S., & Durand-Delacre, D. (2021). Situating 'migration as adaptation' discourse and appraising its relevance to Senegal's development sector. *Environmental Science & Policy, 126*, 11–21. https://doi.org/10.1016/j.envsci.2021.09.008

von Lucke, F. (2021). Principled pragmatism in climate policy? The EU and changing practices of climate justice. *Political Geography, 86*, 102355. https://doi.org/10.1016/j.polgeo.2021.102355

Mager, T. (2018). *The emergency Trust Fund for Africa: Examining methods and motives in the EU'S external migration agenda* [UNU-CRIS policy briefs]. UNU Institute on Comparative Regional Integration Studies.

Malm, A. (2016). *Fossil capital: The rise of steam-power and the roots of global warming*. Verso.

Manghi, S. (2004). *La conoscenza ecologica: Attualità di Gregory Bateson*. R. Cortina.

Mayer, B., & Cournil, C. (2016). Climate change,migration and human rights: Towards group-specific protection? In O. Quirico & M. Boumghar (Eds.), Climate change and human rights: An international and comparative law perspective (pp. 188–203). Routledge.

Mazzara, F. (2019). *Reframing migration: Lampedusa, border spectacle and the aesthetics of subversion* (P. Antonello & R. S. C. Gordon, Eds.; 1st, new ed ed.). Hachette livre. https://doi.org/10.3726/b15214

Mbembe, A. (2019). *Necropolitics*. Duke University Press.

McLeman, R. (2018). Thresholds in climate migration. *Population and Environment, 39*(4), 319–338. https://doi.org/10.1007/s11111-017-0290-2

Membretti, A., Barbera, F., & Tartari, G. (2024). Migrazioni verticali: La montagna ci salverà?. Donzelli editore.

Mezzadra, S. (2004). *The Right to Escape. Ephemera, 4*(3), 267–285.

Miller, T., Buxton, N., & Akkerman, M. (2021). Global climate wall: How the world's wealthiest nations prioritse borders over climate action. Transnational Institute. https://www.tni.org/en/publication/global-climate-wall

Moore, J. W. (2017). *Antropocene o capitalocene?* Scenari di ecologia-mondo nella crisi planetaria.

Musarò, P. (2017). Mare nostrum: The visual politics of a military-humanitarian operation in the Mediterranean Sea. *Media, Culture & Society, 39*(1), 11–28. https://doi.org/10.1177/0163443716672296

Musarò, P., Papastergiadis, N., & Peja, L. (2022). Reimagining narratives on migration. The role of media, arts and culture in promoting transcultural dialogue. *Comunicazioni Sociali, 1*, 3–16. https://doi.org/10.26350/001200_000154

Naser, M. M., Swapan, M. S. H., Ahsan, R., Afroz, T., & Ahmed, S. (2019). Climate change, migration and human rights in Bangladesh: Perspectives on governance. *Asia Pacific Viewpoint, 60*(2), 175–190. https://doi.org/10.1111/apv.12236

Neef, A., Pauli, N., & Salami, B. (Eds.). (2024). *De Gruyter handbook of climate migration and climate mobility justice*. De Gruyter. https://doi.org/10.1515/9783110752144

Neef, A., & Sheller, M. (2024). 22 towards climate mobility justice: From climate debt to climate reparations and a welcoming culture for climate migrants. In A. Neef, N. Pauli, & B. Salami (Eds.), *De Gruyter handbook of climate migration and climate mobility justice* (pp. 465–476). De Gruyter. https://doi.org/10.1515/9783110752144-023

Neocleous, M. (2019). Securitati perpetuae: Death, fear and the history of insecurity. *Radical Philosophy, 206*, 19–33. https://www.radicalphilosophy.com/article/securitati-perpetuae

Neu, F. N., & Fünfgeld, H. (2022). Political ecologies of resettlement in river deltas. *Geography Compass, 16*(6), e12621. https://doi.org/10.1111/gec3.12621

Nshimbi, C. C., & Moyo, I. (2016). Visible and invisible bordering practices: The EU-African migration conundrum and spatial mobility of borders. *World Journal of Science, Technology and Sustainable Development, 13*(4), 300–314. https://doi.org/10.1108/WJSTSD-01-2016-0014

Nyantakyi-Frimpong, H., & Dinko, D. H. (2022). Climate and non-climatic stressors, internal migration, and belonging in Ghana. In T. Mayer & T. Tran (Eds.), *Displacement, belonging, and migrant Agency in the Face of power* (1st ed., pp. 153–165). Routledge. https://doi.org/10.4324/9781003170686

Owens, C. H., Lee, M. J., Grim, M., Schroeder, J., & Young, H. S. (2023). Interactions between temperature and predation impact insect emergence in alpine lakes. *Ecosphere, 14*(7), e4619. https://doi.org/10.1002/ecs2.4619

Palinkas, L. A. (2020). Global climate change, population displacement, and public health: The next wave of migration. *Springer International Publishing*. https://doi.org/10.1007/978-3-030-41890-8

Papastergiadis, N. (2023). *The cosmos in cosmopolitanism*. Polity Press.

Paprocki, K. (2020). The climate change of your desires: Climate migration and imaginaries of urban and rural climate futures. *Environment and Planning D: Society and Space, 38*(2), 248–266. https://doi.org/10.1177/0263775819892600

Parenti, C., & Moore, J. W. (Eds.). (2016). *Anthropocene or capitalocene? Nature, history, and the crisis of capitalism*. PM Press.

Parsons, L. (2019). Structuring the emotional landscape of climate change migration: Towards climate mobilities in geography. *Progress in Human Geography, 43*(4), 670–690. https://doi.org/10.1177/0309132518781011

Pellizzoni, L. (2021). Antropocene come dispositivo governamentale. *Politica & Società, 3*, 495–505. https://doi.org/10.4476/103014

Piguet, E. (2013). From "primitive migration" to "climate refugees": The curious fate of the natural environment in migration studies. *Annals of the Association of American Geographers, 103*(1), 148–162. https://doi.org/10.1080/00045608.2012.696233

Piguet, E. (2022). Linking climate change, environmental degradation, and migration: An update after 10 years. *WIREs. Climate Change, 13*(1). https://doi.org/10.1002/wcc.746

Povinelli, E. A. (2021). Between Gaia and Ground: Four axioms of existence and the ancestral catastrophe of late liberalism. *Duke University Press*. https://doi.org/10.1215/9781478021872

Quijano, A. (2000). Coloniality of power and eurocentrism in Latin America. *International Sociology, 15*(2), 215–232. https://doi.org/10.1177/0268580900015002005

Randall, A. (2017). Engaging the media on climate-linked migration. In B. Mayer & F. Crépeau (Eds.), *Research handbook on climate change, migration and the law*. Edward Elgar Publishing. https://doi.org/10.4337/9781785366598.00021

Ranganathan, M., & Bratman, E. (2021). From urban resilience to abolitionist climate justice in Washington. *DC. Antipode, 53*(1), 115–137. https://doi.org/10.1111/anti.12555

Ransan-Cooper, H., Farbotko, C., McNamara, K. E., Thornton, F., & Chevalier, E. (2015). Being(s) framed: The means and ends of framing environmental migrants. *Global Environmental Change, 35*, 106–115. https://doi.org/10.1016/j.gloenvcha.2015.07.013

Raty, T., & Shilhav, R. (2020). The EU Trust Fund for Africa: Trapped between aid policy and migration politics. *Oxfam*. https://doi.org/10.21201/2020.5532

Renou, Y., & Diallo, A. (2019). Changement climatique et migrations: Qualification d'un problème, structuration d'un champ scientifique et activation de politiques publiques. *Mondes En Développement, 4*(172), 87–107.

Rose, N. (2012). *Migration als Bildungsherausforderung: Subjektivierung und Diskriminierung im Spiegel von Migrationsbiographien* (Vol. 29, 1st ed.). transcript Verlag. https://doi.org/10.14361/transcript.9783839421352

Russo, K. E., & Wodak, R. (2019). Introduction: The representation of "exceptional migrants" in media discourse. The case of climate-induced migration. *Anglistica AION an Interdisciplinary Journal, 21*(2), 1–6. https://doi.org/10.19231/angl-aion.201720

Rychnovskà, D. (2014). Securitization and the power of threat framing. *Perspectives, 22*(2), 9–31.

Sahraoui, N. (2023). Étudier les facteurs environnementaux des migrations contemporaines au prisme du genre. In A. Lenoël & J. Valette (Eds.), *Dossier «Migrations et climat: La fonte des certitudes»*. De facto.

Sakellari, M. (2024). Communicating climate change induced migration: The role of NGOs. *Open Research Europe, 3*, 163. https://doi.org/10.12688/openreseurope.16232.2

Schäfer, M. S., Scheffran, J., & Penniket, L. (2016). Securitization of media reporting on climate change? A cross-national analysis in nine countries. *Security Dialogue, 47*(1), 76–96. https://doi.org/10.1177/0967010615600915

Schewel, K. (2020). Understanding immobility: Moving beyond the mobility bias in migration studies. *International Migration Review, 54*(2), 328–355. https://doi.org/10.1177/0197918319831952

Scissa, C. (2024). Human mobility in the context of disasters, climate change and environmental degradation in the Euro-Mediterranean region: Challenges and insights.

Scott, M. (2023). Adapting to climate-related human mobility into Europe: Between the protection agenda and the deterrence paradigm, or beyond? *European Journal of Migration and Law, 25*(1), 54–82. https://doi.org/10.1163/15718166-12340144

Servigne, P., & Stevens, Raphaël. (with Cochet, Y.). (2021). *Convivere con la catastrofe: Piccolo manuale di collassologia* (S. Bertolini, Trans.). Treccani.

Sheller, M. (2018). *Mobility justice: The politics of movement in the age of extremes*. Verso.

Sheller, M. (2020). Mobility justice in urban studies. In O. B. Jensen, C. Lassen, V. Kaufmann, M. Freudendal-Pedersen, & I. S. Gøtzsche Lange (Eds.), *Handbook of urban Mobilities* (1st ed.). Routledge. https://doi.org/10.4324/9781351058759

Sheller, M., & Urry, J. (2006). The new Mobilities paradigm. *Environment and Planning A: Economy and Space, 38*(2), 207–226. https://doi.org/10.1068/a37268

Solórzano, D. G., & Yosso, T. J. (2002). Critical race methodology: Counter-storytelling as an analytical framework for education research. Qualitative Inquiry, 8. (1), Article 1 doi:https://doi.org/10.1177/107780040200800103.

Spivak, G. C. (1988). Can the Subaltern Speak? In C. Nelson, & L. Grossberg (Eds.), Marxism and the Interpretation of Culture. University of Illinois Press.

Squire, V. (Ed.). (2012). *The contested politics of mobility: Borderzones and irregularity (first issued in paperback)*. Routledge Taylor & Francis Group.

Tazzioli, M. (2020). The making of migration: The biopolitics of mobility at Europe's Borders. SAGE Publications, doi:https://doi.org/10.4135/9781526492920.

Telford, A. (2018). A threat to climate-secure European futures? Exploring racial logics and climate-induced migration in US and EU climate security discourses. *Geoforum, 96*, 268–277. https://doi.org/10.1016/j.geoforum.2018.08.021

Transnational Institute. (2021). *Annual Report*, 2021. https://www.tni.org/en/publication/annual-report-2021

Tschakert, P., & Neef, A. (2022). Tracking local and regional climate im/mobilities through a multidimensional lens. *Regional Environmental Change, 22*(3), 95. https://doi.org/10.1007/s10113-022-01948-6

Turton, D. (2003). *Conceptualising forced migration (refugee studies Centre working papers series 13)*. Oxford University.

Van Aken, M. (2020). *Campati per aria*. Elèuthera.

Vergès, F. (2017). Racial Capitalocene. In G. T. Johnson & A. Lubin (Eds.), *Futures of Black radicalism* (pp. 72–82). Verso.

Vince, G. (2022). *Nomad century: How climate migration will reshape our world* (first U.S. edition). Flatiron Books.

Warner, J., & Boas, I. (2017). SECURITISATION of climate change: THE risk of exaggeration. *Ambiente & Sociedade, 20*(3), 203–224. https://doi.org/10.1590/1809-4422asocex0003v2022017

Watson, S. (2011). The 'human' as referent object? *Humanitarianism as securitization. Security Dialogue, 42*(1), 3–20. https://doi.org/10.1177/0967010610393549

Wiegel, H., Boas, I., & Warner, J. (2019). A mobilities perspective on migration in the context of environmental change. *WIREs Climate Change, 10*(6), e610. https://doi.org/10.1002/wcc.610

Wilkinson, E., Kirbyshire, A., Mayhew, L., Batra, P., & Milan, A. (2016). *Climate-induced migration and displacement: Closing the policy gap*. Overseas Development Institute (ODI). https://environmentalmigration.iom.int/sites/g/files/tmzbdl1411/files/documents/2023-11/10996.pdf

Yumagulova, L., Parsons, M., Woman-Munro, Y. O., & D., Dicken, E., Lambert, S., Vergustina, N., Scott, J. C., Michell, P., & Black, W. (2023). Indigenous perspectives on climate mobility justice and displacement-mobility-immobility continuum. *Climate and Development, 1–18*. https://doi.org/10.1080/17565529.2023.2227158

Yusoff, K. (2018). *A billion black Anthropocenes or none*. University of Minnesota Press.

Zardo, F. (2022). The EU Trust Fund for Africa: Geopolitical space making through migration policy instruments. *Geopolitics, 27*(2), 584–603. https://doi.org/10.1080/14650045.2020.1815712

CHAPTER 5

Awareness Campaigns on Climate (Im)Mobilities

Introduction

Information, communication, and awareness campaigns—tools designed to direct individual behavior toward socially desirable outcomes (Weiss & Tschirhart, 1994)—are employed across various domains of societal concern, including public health, education, criminal justice, environmental awareness, equal opportunity, and combating gender and ethnic discrimination.

Since the 1990s, such campaigns have been increasingly utilized as part of migration management and control strategies (Nieuwenhuys & Pécoud, 2007). This approach aligns with broader practices of securitizing national borders and externalizing migration policies, which have evolved from the late twentieth century into the twenty-first as mechanisms to govern the mobility of those deemed 'undesirable' by the Global North (Cappi & Musarò, 2022). Initially targeting populations in migrant-sending regions such as Central and Eastern Europe, South-East Asia, and Central America, these campaigns shifted focus after 2000 to countries in Northern and Sub-Saharan Africa (e.g. Morocco, Gambia, Ghana, Senegal, and Nigeria), as well as regions in the Middle East (e.g. Afghanistan, Iran, Iraq) and Sri Lanka (Schloenhardt & Philipson, 2013). Ostensibly aimed at raising awareness about the risks associated with irregular migration—such as human trafficking dangers, challenges faced by

E. Giacomelli and P. Musarò, *Climate Mobility Justice*,
https://doi.org/10.1007/978-3-032-07907-7_5

returnees, and the difficulties awaiting migrants in destination countries—these campaigns rest on the assumption that irregular migration is largely driven by misinformation or ignorance. Accordingly, they advocate for providing accurate information as a means to reshape perceptions and influence the decision-making processes of prospective migrants (Fiedler, 2020; Pécoud, 2010; Schans & Optekamp, 2016). By promoting immobility as a 'safe' alternative, migration awareness campaigns often obscure structural inequalities in sending countries and legitimize the externalization of border control, effectively extending state power beyond sovereign territories. In doing so, they become performative acts that prioritize securitization over addressing the root causes of displacement—ultimately serving domestic political agendas rather than migrant welfare.

In parallel, research on climate change awareness campaigns has also evolved, reflecting increasingly diverse approaches to public engagement on this pressing global issue. Early efforts, such as the European Commission's 2006 Climate Change Awareness Campaign, investigated how ethical consumption could be promoted through persuasive communication, with studies like Manuti (2013) analyzing its discursive impact on behavioral change among university students.

Recent initiatives have expanded the scope of climate communication by integrating participatory methods and co-creation processes. For instance, Dziminska et al. (2021) analyzed citizen proposals to improve climate change communication through thematic analysis, advocating for community involvement as a key to fostering pro-environmental behavior. Other comprehensive initiatives, such as those cataloged by *FundsforNGOs,* promote multimedia campaigns and capacity-building programs to empower a wide range of stakeholders—from students to policymakers—in climate action.

While these examples point to the growing sophistication of awareness campaigns, they often fail to account for the interconnections between climate change and migration narratives. Scholarly inquiries into these phenomena have largely run on parallel tracks, with awareness campaigns often siloing climate change and migration as separate crises. Early work on migration deterrence, such as Kosnick's (2014) media archaeology, traced the ways security paradigms have shaped migration narratives. Bishop (2020) analyzed asylum-seeker media campaigns by narrowly focusing on geopolitical borders. Brändle's study (2022) examined EU digital campaigns, finding that state-sponsored messaging often breeds distrust while arguing for culturally resonant, peer-to-peer networks to counter misinformation. These campaigns tend to narrate migration either as a crisis or solution, as a threat or tragedy, oscillating between humanitarian appeals and securitarian logics. Cappi and Musarò (2022) analyze awareness campaigns targeting potential migrants as instruments of symbolic bordering, arguing they reinforce neoliberal governance by discouraging mobility through fear-based narratives. Campaigns like Australia's 'No Way' and Italy's 'Aware Migrants,' frame migration as perilous and futile, emphasizing risks like financial ruin, familial suffering, and death, while portraying migrants as gullible victims of smugglers' deceit.

The emerging focus on climate (im)mobilities seeks to bridge this gap by addressing the interplay between climate change and (im)mobilites. This interdisciplinary shift underscores the need for campaigns that integrate climate change and migration into cohesive, mutually informing frameworks of public engagement.

Beneath the flickering glow of our screens—where the mediapolis (Silverstone, 2009) unfolds as a digital amphitheater of modern consciousness, information, communication, and awareness—campaigns on climate (im)mobility function as both lanterns and labyrinths. They shine a light on the crises we face while entangling them in webs of contested meaning, with pixels and policy briefs merging into symbolic tools of governance. Curated by states, NGOs, and intergovernmental entities, these campaigns do more than just convey information—they mold social imaginaries, fluctuating between urgency and containment. Their power lies as much in what they reveal as in what they conceal as they map moral geographies that reinforce borders as both material and symbolic tools of exclusion (Musarò, 2019).

Recent scholarship has shed critical light on how these campaigns shape perceptions of climate (im)mobility. Giacomelli and Cappi (2025), for instance, show how fear-based narratives of intergovernmental and NGO campaigns reproduce a Panicocene sensibility, depicting climate migrants as destabilizing 'others.' Similarly, Sakellari's (2024) examination of UK-based NGOs highlights how visual and textual narratives depoliticize migrants, emphasizing their 'propensity to bring social instability' while neglecting systemic drivers of displacement.

As discursive technologies, these awareness campaigns operate within what Foucault (1971) calls the 'order of discourse': a space where language creates and legitimizes certain truths while silencing others. Defined as 'purposive attempts to inform, persuade, or motivate behavior changes in a relatively well-defined and large audience, generally for non-commercial benefits to the individuals and/or society at large, typically within a given time period, by means of organized communication activities' (Rice & Atkin, 2013, p. 16), these campaigns are epistemic tools that frame climate (im)mobility through curated words like 'resilience,' 'crisis,' 'vulnerability.' Each term, as Fairclough (2001) reminds us, is itself a social practice, subtly embedding hierarchies of risk and worth. A flooded village becomes a humanitarian vignette; a displaced family morphs into a statistical 'flow.' Such representations, widely circulated across social media and institutional reports, do not merely describe reality: by amplifying certain truths while silencing others, these campaigns construct what Buraschi and Aguilar Idáñez (2016) call 'moral boundaries' that delineate who deserves care and who warrants control; who is seen as a victim and who is cast as a threat. Through their visual grammar—images of flooded landscapes or parched earth—and their lexical choices—terms like 'flows' or 'resilience'—these campaigns aestheticize climate change while depoliticizing its causes. They contribute to transforming structural violence into a natural disaster, erasing histories of exploitation that have fractured ecosystems and displaced communities, and thus prefiguring material borders by shaping public consent for exclusionary policies.

Herein lies the paradox: campaigns designed to foster solidarity often end up replicating the very logic they seek to overcome. Commonly hailed as tools of enlightenment, these campaigns are far more than neutral instruments of communication, they are deeply political acts of framing, bordering, and worldmaking. As nation-states, intergovernmental bodies, and NGOs deploy these campaigns to address environmental degradation and the specter of 'uncontrolled' migration, they do not merely describe reality; they actively construct it. By mobilizing securitized tropes of 'uncontrolled migration' alongside appeals for climate justice, they straddle the humanitarian-security nexus (Pécoud, 2010), a discursive tightrope that prefigures the 'soft borders' of public perceptions (Musarò, 2019).

In line with these reflections, this chapter interrogates awareness campaigns on climate (im)mobilities as technologies of visibility, asking not only what they show but what they foreclose. Moving across social markers, it traces how their narratives either reinforce or rupture the us/them dichotomies underpinning contemporary border regimes. Are these campaigns just 'blaming nature' or do they unmask injustices? Drawing on Musarò's analysis (2019) of how awareness campaigns function as symbolic bordering practices that extend state power beyond territorial borders—a critique of awareness as borderwork—it explores how these campaigns, even in their most progressive iterations, risk reproducing the epistemic violence of universalizing Western climatologies and mobility imaginaries.

What stories do we tell when migration meets the climate crisis? What narratives do awareness campaigns weave as they seek to inform, persuade, and mobilize? Do these campaigns reflect the dichotomies of care and control that underpin border governance? To address these questions, this chapter analyzes the following campaigns through decolonial, intersectional, and feminist frameworks: Do the Right Thing (IOM); Together, We Can #FindAWay (IOM); Think of Tomorrow, Act Today (IOM); Climate Refugees (Climate Refugees); Climate Justice is Migrant Justice (Joint Council for the Welfare of Immigrants -JCWI); Climate: All Migrants Tomorrow? (Agence Française de Développement Group—AFD); Climate and Migration Coalition (Climate and Migration Coalition); the Pacific Islands Students Fighting Climate Change (PISFCC). To study these campaigns is to map out the contours of a fraught terrain: one where every infographic is a micro-border, every testimony a site of resistance. Their power resides not in their reach but in their capacity for worldmaking—a power that is as fragile as it is formidable.

The Power of Representation

The information campaigns analyzed in this chapter aim to illuminate the intricate and multifaceted relationship between climate change and migration, seeking not only to inform but also to reimagine public understanding of these defining phenomena of the twenty-first century. Conceived and implemented by international institutions and NGOs, these campaigns weave narrative and visual rhetoric to transform collective imaginaries, bridging the abstract and the tangible. Although it is too early to fully assess their efficacy in raising awareness, fostering sensitivity, and shaping attitudes or behaviors (Browne, 2015), it is crucial to interrogate the motivations, frames, causes, and narratives they employ. Such an inquiry offers a preliminary glimpse into the intended—and unintended—consequences these campaigns may set in motion.

At their core, communication campaigns engage with public opinion, understood as a dynamic and collective expression of societal attitudes at a given time. We know that images can have significant influence;

however, they are often dehumanizing and problematic for a variety of reasons, including issues of consent and depictions of distress and victimization. Moreover, these images frequently rely on familiar visual clichés, which oversimplify complex issues and risk perpetuating harmful visual stereotypes. For a campaign addressing environmental migration to effect meaningful change, it must navigate three interconnected dynamics: perceptions shape attitudes; attitudes drive behavior; and behavior, in turn, reshapes perceptions. The key challenge lies in crafting narratives that are evocative but not alarmist—informative without inducing crippling fear or apathy.

The design, implementation, and evaluation of communication campaigns on environmental migration vary widely across contexts, reflecting the complexity of the issue. The campaigns examined here acknowledge that long-term impact requires more than information dissemination; it demands active engagement with target audiences as co-creators of meaning. Communication on such issues must be dialogical rather than unidirectional, fostering a reciprocal exchange of perspectives. This imperative is further complicated by the ever-evolving 'medi-apolis'—a global media landscape shaped by accelerating connectivity and technological innovation. As a result, campaign strategies are diverse, often experimental, and increasingly non-traditional. Dissemination channels span both digital and physical spaces: brochures, posters, stickers, television programs, radio broadcasts, films, advertisements, social media platforms, and digital posts are complemented by grassroots initiatives such as live community events, public conferences, workshops, and capacity-building activities. Some campaigns also leverage the cultural capital of influencers and celebrities to amplify their reach.

Nevertheless, these efforts are not without significant challenges. Chief among them is the widespread lack of public knowledge regarding environmental migration—a gap compounded by contextual complexities and communication limitations. Another common shortcoming is the absence of clearly defined objectives. A central question for any campaign must be: what change do we seek to inspire? A well-conceived campaign requires a nuanced analysis of the issue at hand, strategic framing of its message, precise identification of target audiences, and a shared vision of a desired future.

Equally important are considerations regarding the specific actions to be taken, the stakeholders involved, the resources available (both internal and external), the tactics deployed, and the timing of the interventions. In the context of environmental migration, strategies must take into account the potent imaginaries evoked by terms like 'environmental migrants' or 'climate refugees.' Though undefined in international law, these terms carry profound symbolic weight—shaping perceptions through emotional resonances that can either humanize or alienate. Navigating this terrain requires balancing urgency with empathy to promote understanding while avoiding reductive or alarmist framings.

Awareness campaigns can spotlight the link between environmental harm and migration, expose power dynamics fueling today's crises, and help drive meaningful change in how we address climate mobility. For this reason, we consider it crucial that they exercise respons-ability in responding to the 'toxic' narratives that combine the most negative traits of both climate and migration discourse. In recent years, for example, it has become common to claim that climate change will trigger massive shifts of 'climate refugees' from the Global South to the Global North. Among others, António Guterres, secretary-general of the United Nations, warned of 'a mass exodus' of this kind on a 'biblical scale,' adding that this migration could unleash a 'torrent of trouble.'[1]

The idea of a 'mass exodus' evokes images of enormous, sudden movements of people. Words like 'mass,' 'biblical,' 'exodus,' and 'torrent' tap into deep-seated fears and moral panics surrounding apocalyptic scenarios of climate-induced migration. These frames flatten the complexity of the discourse and speak most directly to audiences in the Global North already predisposed to hostility toward migrants and refugees. This resulting imaginary often features images of desperate masses crossing borders on foot.

However, the reality of climate-induced migration is quite different. In relation to slow-onset environmental phenomena, many people engage in temporary and seasonal migration when farming becomes less viable, typically relocating from rural areas to nearby cities in search of work. Most often they use migration as a temporary measure, with movement based on alternating droughts. In sudden weather-related disasters—like typhoons or floods—people generally move along the shortest route to safety. While millions are indeed displaced, many return home after the disaster to rebuild. Others use existing legal migration channels—such as work or education visas—to escape worsening conditions. Conversely,

[1] www.theguardian.com/environment/2023/feb/14/rising-seas-threaten-mass-exodus-on-abiblical-scale-un-chief-warns.

climate change also entraps people and can reduce migration, especially when climate extremes erode livelihoods, leaving people without the resources required to move. Indeed, there are always multiple factors that determine a person's displacement. Displacement is almost never caused by climate change alone but results from a combination of factors, including ongoing conflicts, histories of colonization and oppression, and a lack of infrastructure investment.

Apocalyptic migration predictions, when used to spur climate action, are not only intellectually dishonest but also dangerous. They distort public understanding and weaken the case for effective responses (de Hass, 2020). These narratives, then, have the power to shape public perception and subsequent policy responses to climate-induced migration in potentially disastrous ways. For example, Raimi et al. (2024) conducted an analysis examining the impact of climate migration communication on public perceptions in the United States. Their findings reveal that exposure to narratives concerning cross-border climate migration does not significantly alter public support for climate change mitigation policies. Conversely, narratives focusing on internal climate migration within the United States have been shown to increase support for such policies. While cross-border climate migration narratives did not influence attitudes toward climate policy, they certainly heightened the perception of migration as a risk compared to narratives centered on internal displacement. These findings suggest that emphasizing dramatic future migration across borders may heighten public anxiety rather than mobilize support, while communication about internal displacement narratives may offer a more effective path toward support for climate action. The following analysis explores how the selected campaigns reflect—or counter—these toxic framings.

The Campaigns Analyzed

Alongside media and human rights professionals, INGOs, the UN, and other organizations play a key role in shaping narratives, influencing behavior, amplifying vulnerable voices, and advocating for policy change. This study examines eight communication campaigns conducted between 2020 and 2025, each designed to enhance awareness and provide information on the intricate relationship between climate change and (im)mobilities. Three of these campaigns were spearheaded by the International Organization for Migration (IOM), while the remaining five were implemented by international nongovernmental organizations (INGOs). This selection enables a comparative analysis of the narratives promoted by organizations operating at different institutional levels to promote public understanding of climate-induced (im)mobilities.

These information, communication, and awareness campaigns serve as prismatic case studies—texts and images that can be dissected as sites where 'language becomes a cartography of power' (Baldwin & Bettini, 2017). Both through what they say and what they omit, these campaigns wield the power to define the moral landscape of climate (im)mobility. The campaigns examined are as follows:

- The 'Climate Justice is Migrant Justice'[2] campaign, led by the Joint Council for the Welfare of Immigrants (JCWI), emphasizes the deep interconnection between climate justice and migrant justice movements. This initiative advocates for systemic change by challenging hostile immigration policies that prevent displaced individuals from seeking safety and by promoting climate reparations to address the disproportionate harm caused by fossil fuel industries. Rooted in anti-racism, the campaign centers the voices of migrants in calls for a fairer, greener planet. Through legal advocacy, policy proposals, and public messaging, JCWI seeks counter-narratives of fear and division while fostering solidarity and resistance. By foregrounding the impacts of climate change on marginalized communities, the campaign aims to shift public discourse on migration and climate breakdown toward a justice and inclusion-based framework.[3] Situated at the intersection between climate and migrant justice movements, it powerfully confronts the securitization and violence of borders as a misguided response to climate displacement.[4]
- The 'Climate: All Migrants Tomorrow?'[5] campaign, launched by the Agence Française de Développement (AFD) Group, addresses the growing phenomenon of climate-induced migration through a

[2] https://jcwi.org.uk/campaigns/climate-justice-is-migrant-justice/.

[3] November 22nd, 2024 will mark the first Climate and Migrant Justice Day, as the Joint Council for the Welfare of Immigrants (JCWI) demands support for people affected by climate destruction and calls for an end to border violence.

[4] https://jcwi.org.uk/campaigns/climate-justice-is-migrant-justice/protect-people-not-polluters/.

[5] https://www.afd.fr/en/climate-all-migrants-tomorrow.

three-pronged strategy. First, it focuses on preventing displacement by supporting public policies for climate mitigation and adaptation, financing ecosystem restoration, and reducing exposure to disasters. Second, it enhances response mechanisms to climatic hazards by mobilizing resources for immediate action following major events. Finally, the campaign supports migrant populations through financial transfers that enable communities to develop localized adaptation strategies. Emphasizing both prevention and support, AFD's initiative highlights the need for integrated approaches to climate (im)mobility while advocating for clearer legal recognition of climate migrants and refugees. Through these efforts, the campaign seeks to foster resilience among vulnerable populations while addressing systemic inequalities exacerbated by climate change.

- The 'Climate and Migration Coalition' campaign[6] champions the rights and welfare of those displaced by climate change, advocating for humane and empowering policies that prioritize dignity and survival. This initiative foregrounds two key principles: the 'right to stay'—helping communities thrive at home despite environmental challenges—and the 'right to move,'—promoting safe and legal pathways for migration when staying in place is no longer viable. Through policy advocacy, public engagement, and storytelling, the Coalition seeks to shift public discourse away from securitized narratives of climate migration toward justice-centered approaches. By amplifying the voices of those directly affected, the campaign fosters informed debates on climate-linked migration while addressing systemic gaps in international protections. Its work underscores the need for solidarity-driven responses that recognize migration as a legitimate adaptation strategy in the face of climate breakdown. Their goal is to ensure people-centered policy responses at national and international levels. It specifically supports policies that enable migrants to strengthen their survival capacity and demands adequate assistance and protection for those displaced both within and across borders due to either slow or sudden-onset disasters.[7]

[6] https://climatemigration.org.uk/.

[7] https://climatemigration.org.uk/our-work/.

- The Pacific Islands Students Fighting Climate Change (PISFCC)[8] campaign stands as a powerful testament to youth-led advocacy for climate justice in the Pacific. Founded in 2019 by 27 law students from eight Pacific Island nations, this initiative seeks to address the existential threat of climate change by leveraging international legal mechanisms. The campaign's core objective is to secure an Advisory Opinion from the International Court of Justice (ICJ) clarifying state obligations under international law to protect the rights of present and future generations from the adverse impacts of climate change. Through grassroots mobilization, PISFCC engages Pacific youth across all educational levels, empowering them to participate in global fora, such as COP27 and COP26. The campaign also emphasizes intergenerational equity and human rights, challenging world leaders to recognize the disproportionate burden placed on Pacific Island communities. By combining legal advocacy with public outreach, PISFCC aims both to ensure the survival of Pacific homelands and to foster international solidarity around climate-induced displacement and ecological justice. In the last years, the Pacific Islands Students Fighting Climate Change (PISFCC) has achieved a significant court win at the International Court of Justice (ICJ). This ruling, which clarifies countries' legal obligations to limit climate change and provide restitution for harm caused, was spearheaded by this group of young activists from Pacific island nations. This ruling is a landmark moment for climate justice, particularly for vulnerable island nations facing the brunt of climate impacts.[9]
- Climate Refugees[10] is a non-profit organization dedicated to raising awareness of the human rights and humanitarian dimensions of climate-induced displacement. Using personal stories, visual narratives, and expert insights to evoke emotional engagement within civil society, it aims to mobilize movements advocating for policy change at both national and international levels.

[8] https://www.pisfcc.org/.

[9] More information about this climate case available at: https://www.bbc.com/future/article/20250721-how-a-bold-student-plan-swept-to-un-world-court; https://www.theguardian.com/world/2025/jul/25/pacific-students-who-won-climate-case-icj-international-court-of-justice-hague.

[10] www.climate-refugees.org/storytelling.

- 'Do the Right Thing,'[11] funded by the IOM Development Fund and launched in 2021, sought to highlight the interconnection between migration, the environment, and climate change. A primarily digital campaign, it was also accompanied by an exhibition in Geneva and a music festival in Switzerland. According to IOM, the campaign used various media formats—portraits, quotes, and videos accessible through QR codes—to amplify the voices of migrants, youth, and other underrepresented groups. It showcased projects in Haiti, the Marshall Islands, Niger, Peru, and Tajikistan. While its target audience was not explicitly defined, the campaign appears to have been primarily aimed at engaging international donors interested in addressing climate change and migration.
- 'Think of Tomorrow, Act Today,'[12] released for the COP28 in Dubai (2023), calls for concrete actions to address climate change and its impacts on migration. Attesting to the urgency of addressing the two issues together, IOM launched a forward-looking campaign that includes a website with immersive 360° videos that capture viewers with a sensorially compelling experience. These videos help stakeholders better understand the challenges of climate change and migration, encouraging the public to actively support the cause. The campaign also includes a paper[13] with updated data and a series of actions for a comprehensive response meant to harness migration as a force for sustainable development.
- 'Find A Way',[14] launched in 2020 with unearmarked funding from IOM, is an ongoing communication and fundraising effort. It invites public support for climate migrants, emphasizing their resilience and agency amidst displacement caused by climate change. Highlighting projects in Madagascar, Mongolia, Myanmar, Micronesia, and the Bahamas, the campaign seeks to engage a broad audience, from civil society organizations and governments to intergovernmental organizations, journalists, and private sector actors.

[11] https://www.iom.int/iom-development-fund.

[12] https://acttoday.iom.int/.

[13] https://publications.iom.int/books/thinking-about-tomorrow-acting-today-future-climate-mobility.

[14] https://findaway.iom.int/en/.

Methods

This study employs a critical approach, weaving together discourse analysis and visual semiotics to unravel the narrative architectures shaping climate mobilities. Grounded in Foucault's (1980) conception of power-knowledge regimes and Van Dijk's (1994) socio-political critique of textual hegemony, our methodology interrogates how institutional actors script climatic displacement through multimodal storytelling.

In terms of campaign design, we reference toolkits developed by IOM[15] (2020), Amnesty International[16] (1997), the Communications Campaign Best Practices[17] (Movement Advancement Project, Gay & Lesbian Alliance Against Defamation, 2008), and Climate Outreach's theory of change[18] (Clarke et al., 2020), which outline key principles for building an effective campaign.

Drawing on the analytical frameworks of Boas et al. (2022) for climate (im)mobilities and Giacomelli and Walker (2025) for participatory action research on climate diaries, our analysis moves beyond content evaluation. We map the semiotic ecosystems of each campaign—websites, audiovisual materials, community toolkits—through a decolonial lens that challenges Eurocentric epistemologies (Udah, 2024). This tripartite analytical matrix examines:

- **Discursive Terrains**: How humanitarian logics intersect with securitization tropes (i.e., portraying 'climate refugees' as threat multipliers), and whether resilience narratives risk (however inadvertently) neoliberalizing the practice of adaptation (Methmann & Oels, 2015).
- **Representational Optics**: The visual economy of migrant iconography—do images perpetuate 'vulnerability porn' (Climate Outreach, 2024), or do they subvert victimhood through depictions of agency?
- **Intervention Architectures**: Campaign efficacy metrics against Browne's (2015) typology of behavioral change, especially the role

[15] https://publications.iom.int/books/iom-public-communication-campaign-toolkit.

[16] www.amnesty.org/en/documents/act10/002/2001/en/.

[17] www.lgbtmap.org/communications-campaign-best-practices.

[18] https://climateoutreach.org/reports/theory-of-change.

of participatory design (e.g., climate circuses, influencer collaborations) in public engagement.

Our multimodal critique adopts DiFrancesco and Young's (2011) 'ecology of signs' paradigm, treating each campaign as a palimpsest where frame[19], text, and image enact dialectical storytelling. Whereas prior studies remained focused on message transmission (Tjaden et al., 2018), we probe the interstitial gaps—the unspoken ideological substrata revealed through absences, primarily through the silenced voices of immobile populations. Framing analysis integrates Ransan-Cooper's (2015) matrix on mobility trajectories to examine whether campaigns replicate apocalyptic 'mass exodus' tropes or center place-based resilience.

Visual methodologies such as those articulated by Kress and van Leeuwen (2010) and Rose (2022), offer tools to decode the semiotic power of imagery in campaigns, such as the use of emotive testimonials in IOM's 'Migrants as Messengers' initiatives to humanize climate migrants. In climate communication, Dupar et al. (2019) stress the importance of localized, solutions-focused storytelling to bridge scientific data and public action, while Harvard's Climate Optimist initiative exemplifies how optimism and tangible co-benefits (e.g., health improvements) drive engagement and support.[20] These studies collectively affirm that ethical, evidence-based campaigns must balance risk awareness and empowerment, leveraging trusted messengers and interdisciplinary tools to navigate the socio-political complexities of climate-driven mobility.

What distinguishes this methodology is its dialogic tension between critique and co-creation. While we deconstruct securitization discourses (Boas, 2015), we also engage Climate Outreach's (2024) call for ethical storytelling—assessing how campaigns can 'humanize without homogenizing' climate displacement. This dual lens reveals the paradox of awareness-raising: the urgency to inform often coexists with the risk of

[19] As observed by Koteyko and Atanasova (2016, p. 6), the identification of different discourses 'has parallels with framing theory and analysis, especially with an approach labeled as a critical approach to researching media frames [...]. This approach views frames in the media as the outcomes of journalists' 'discursive habit' to represent the world in one way instead of another by deploying 'culturally constructed codes that function ideologically, as if they were the creations of nature' (Olausson, 2009, p. 423)'.

[20] https://www.forbes.com/sites/avivahwittenbergcox/2023/04/17/the-power-of-climate-optimists-flip-the-narrative-change-the-future/?sh=3d612fce7d8b.

dystopian framings that paralyze the political imagination (Hartmann, 2010). Through this critical poetics of analysis, we seek to capture not only what is said, but also the spectral presences of what remains unsaid—the narratives that might yet bloom in the fissures of dominant discourses.

The Analysis

Climate (im)mobility campaigns navigate a complex terrain of narratives, encompassing victimhood, securitarian framings, adaptive agency, and the portrayal of nature as an antagonist. This analysis critically engages with distinct lines of inquiry: the victimhood narrative and its scripted vulnerabilities, securitarian framings that cast migrants as threats to stability, the faces of (im)mobilities that foreground self-representation and agency, and the tendency to absolve systemic inequities by blaming nature. It examines how campaigns oscillate between portraying migrants as passive victims of natural disasters and recognizing them as agents of justice and resilience. Through a critical lens on moral geographies, visual rhetoric, and linguistic constructs, this section unpacks how these campaigns navigate the tensions between visibility and erasure, vulnerability and agency—ultimately challenging the binaries that dominate climate (im)mobilities discourse.

The Victimhood Narrative and Scripted Vulnerabilities

The victimhood narrative remains a pervasive yet complex frame in climate (im)mobility campaigns, often shifting between evoking empathy and reinforcing disempowerment. While campaigns such as IOM's *Find A Way*, *Do the Right Thing*, and *Think of Tomorrow, Act Today* aim to humanize displaced populations through individual stories, they frequently depict migrants as passive victims of natural disasters, detaching them from the systemic drivers of their vulnerability. In Madagascar, the campaign's invocation of *kere*—a term describing cyclical drought and famine—captures this tension well. Tisbite Ranaivo's lament, 'The kere is everywhere,' evokes both visceral suffering and an implicit call for containment, framing climate chaos as an omnipresent threat demanding intervention. Notably, *Think of Tomorrow, Act Today* seeks to promote intergenerational justice, portraying youth as central agents of change. By enabling forward-looking and innovative solutions for people to stay, and to be on the move, the campaign outlines why climate mobility is crucial for a common future.

AFD's *Climate: All Migrants Tomorrow?* leans into this victimhood framing while adding a technocratic layer. This campaign emphasizes preventative measures like ecosystem restoration and disaster risk reduction while sidestepping discussions of historical responsibility or systemic injustice. The imagery used—maps of vulnerable regions overlaid with data points—creates urgency but also depersonalizes the human stories behind climate-induced movements. Although the campaign integrates social and ecological dimensions, it refrains from addressing it as a man-made crisis rooted in systemic inequalities.

In contrast, NGO-led initiatives such as JCWI's *Climate Justice is Migrant Justice* and the *Pacific Islands Students Fighting Climate Change* (PISFCC) actively resist victimizing frames by centering justice and agency. JCWI, for instance, directly critiques the victimhood narrative by stating: 'Right now, people in the Global South—who contributed least to the climate crisis—are bearing the brunt of its impact, while the Global North shuts its borders to keep people out.'[21] The campaign refuses nature-blaming tropes and foregrounds systemic injustices—colonial histories, fossil fuel dependency, and economic exploitation—as root causes of displacement. By linking climate breakdown to exploitative economic systems driven by Northern industries, JCWI emphasizes the 'right to stay' and 'right to move.' This campaign amplifies voices like Afghan climate activists denied asylum in the UK, reframing migrants not as objects of pity but as claimants of justice. Likewise, the *Pacific Islands Students Fighting Climate Change* (PISFCC) campaign underscores the political dimensions of climate change and frames it as a human rights issue.

Security: Bordering the Crisis

Securitarian framings recast climate migrants as threats to global stability. The Panicocene thrives on binaries—us/them, threat/victim, savior/saved—a dialectic mirrored in campaigns like IOM and similar intergovernmental agencies, which often mobilize technocratic metaphors like 'flows' to describe migration. Such language, as Boas et al. (2019) note,

[21] https://jcwi.org.uk/campaigns/climate-justice-is-migrant-justice/.

externalizes responsibility and treats mobility as a crisis to be managed, rather than as a response to systemic failure. By contrast, initiatives like JCWI, Climate and Migration Coalition, and PISFCC deliberately avoid such language. They center justice, agency, and rejection of fear-based cartographies. By focusing on demands for climate reparations and the dual 'right to stay/move,' these campaigns reframe migrants as agents of intersectional resistance, weaving climate and border abolition into a single struggle (Chouliaraki, 2024). *Climate Justice is Migrant Justice* (JCWI) is especially explicit in foregrounding both the right to stay and the right to move: 'We believe that, in the face of these challenges, everyone should have the right to stay and the right to move. That means protection from the worst impacts of the climate crisis so that people can stay in their homes and communities—but it also means being able to move across borders to seek safety and a dignified life, if life at home is no longer viable.'[22] Rooted in cosmopolitan solidarity and international justice, the campaign advocates for legal protections that, while certainly upholding the right to move, also support the right to stay by advocating for humane and empowering policies aimed at protecting those most affected by climate change. JCWI's campaign challenges the securitarian framing by advocating for safe routes and legal protections for those displaced by climate breakdown. It explicitly critiques far-right narratives that scapegoat migrants, emphasizing instead that 'everyone should have the right to stay and the right to move.'[23] By centering solidarity and justice over fear and division, JCWI reframes migration as a human right rather than as a security risk. Their demands are unapologetic: 'No climate justice without migrant justice, racial justice, and decolonization.'[24] Their protest visuals juxtapose oil rigs with detention centers, overlaid with slogans like 'Stop Fueling Borders'—a direct challenge to the fossil fuel industry's role in increasing both emissions and border militarization. Insofar as it subverts the topo-logic of Western geopolitics (Brambilla, 2021), the campaign makes visible the 'slow violence' of racial capitalism (Nixon, 2013). Accompanying policy briefs cite Afghani climate activists denied UK asylum, while centering migrant voices as experts: 'We are not victims; we are survivors of your

[22] https://jcwi.org.uk/campaigns/climate-justice-is-migrant-justice/.

[23] https://climatemigration.org.uk/.

[24] https://jcwi.org.uk/campaigns/climate-justice-is-migrant-justice/.

greed.'[25] This aligns with *Climate Outreach*'s principle of 'telling real, detailed stories,' where images of migrant-led climate strikes avoid tropes of helplessness, instead showcasing collective power.[26] Similarly, the *UK Climate & Migration Coalition* warns against security rhetoric that frames migrants as 'invaders,' noting how such narratives amplify anti-immigrant sentiment.[27]

Sakellari (2024) observes that humanitarian and securitarian frames—often opposing 'victims' and 'threats'—are, in fact, interwoven strands of the same moral tapestry. Climate migrants, even when shown as fragile beings through curated imagery of displacement, are often implicitly cast as risks justifying surveillance and control. For example, IOM's Think of Tomorrow, Act Today campaign cites the World Bank's Groundswell Report (Clement et al., 2021), warning 'that over 216 million people could become internal climate migrants by 2050.' The *FindAWay* campaign exemplifies this risk through militarized imagery and alarmist language: in the case study on Mongolia, photographs depict military personnel conducting 'emergency preparedness simulation exercises,' accompanied by captions about monitoring displacement and minimizing irregular migration. Such visuals align with what Sakellari (2024) terms 'security conditioning,' where ecological crises are instrumentalized to justify border fortifications. A caption underscoring 'monitoring displacement' reveals how compassion and securitization coalesce, embedding governmentality into the rhetoric of care. Lexical choices like 'minimizing irregular migration' and 'securing safety' encode urgency into policy, transforming ecological vulnerability into a frontier of regulation (Cappi & Musarò, 2022).

As Ransan-Cooper et al. (2015) argue, campaigns often oscillate between empathy and alarmism, with IOM's materials tending to privilege victimhood and security tropes, while relegating migrant agency to the margins. By contrast, NGOs like Climate Refugees or JCWI resist this frame. They reimagine migrants as political subjects, weaving their stories into a cosmopolitan framework of climate immobilities justice that includes broader struggles for climate reparations and border abolition.

[25] https://jcwi.org.uk/campaigns/climate-justice-is-migrant-justice/.

[26] https://climatevisuals.org/climate-migration/.

[27] https://climatemigration.org.uk/people-are-not-the-problem-why-national-security-is-the-wrong-frame-for-migration-linked-to-climate-change/.

To move beyond these binaries requires adopting decolonial perspectives that foreground intersectional justice and elevate diverse voices without reproducing racialized hierarchies (Tilley et al., 2023). As Hiraide (2022) reminds us, language itself is a terrain of struggle. How we name and frame climate-induced mobility shapes whether solidarity or division prevails. This dance of visibility and erasure—where vulnerability and threat become two sides of the same coin—underscores a central paradox of climate communication: to evoke action, we must first render suffering legible, but legibility also risks reducing the real lives of people into symbols within a larger geopolitical script. The challenge, then, is to craft narratives that transcend this binary and embrace what Butler (2015) calls 'alliances of bodies'—stories in which migrants are neither pawns nor perils, but protagonists in a shared struggle for planetary justice.

Adaptative: Rewriting the Script

Against the backdrop of victimhood and securitarian fears, parts of the analyzed campaigns reframe climate migrants as adaptive agents creatively navigating crises. JCWI takes this framing further by emphasizing migrant agency in both origin and destination contexts. Its steering committee, composed of displaced individuals, develops policy proposals that prioritize dignity and self-determination over paternalistic interventions. This participatory approach positions migrants not merely as survivors or victims but as protagonists of their (possible) futures. The *Pacific Islands Students Fighting Climate Change* (PISFCC) campaign subverts victimhood through legal innovation. Their demand for an ICJ Advisory Opinion reframes adaptation as a rights-based imperative, declaring, 'Climate change now threatens to destroy Pacific homelands and cultures.'[28] Youth-led TEDx talks pair ancestral oceanic navigation charts with excerpts from UN treaty, visually merging Indigenous knowledge and international law. However, while symbolically potent, PISFCC's emphasis leans toward technical solutions rather than structural transformation. AFD's campaign subtly incorporates political subjectivity by spotlighting grassroots leadership in adaptation projects, but it stops short of directly addressing systemic inequities. Similarly, the

[28] https://www.hurights.or.jp/archives/asia-pacific/section1/hreap_v13_experiences6.pdf.

Climate and Migration Coalition's 'right to stay/move' paradigm is illustrated through interactive maps showing both disappearing coastlines and existing migration corridors, thus challenging the binary of 'forced vs. voluntary' movement.[29] These campaigns employ what Climate Visuals terms 'constructive future' imagery[30]—for example, the Guna community receiving keys to newly built homes on mainland Panama, a symbol of collective self-determination amid rising seas. To move beyond 'blaming nature' requires adopting decolonial perspectives that center human agency and systemic accountability. Campaigns like JCWI's and PISFCC's exemplify how narratives can shift from portraying migrants as passive victims to recognizing them as subjects of global justice. Here, adaptive agency emerges not as compliance with containment, but as resistance: a reclaiming of voice in the face of narratives that render mobility either a crisis to fear or a tragedy to pity. By integrating intersectional approaches and amplifying diverse voices, these initiatives challenge reductive binaries and offer transformative pathways for addressing climate-induced mobility.

Faces of (Im)Mobilities

The absence of self-representation in migration communication campaigns has long been placed under scrutiny (Trauner et al., 2024). Running against the grain, the campaigns analyzed here make a concerted effort to include the voices, narratives, and perspectives of local populations. Platforms such as *Climate Refugees* amplify migrant voices through interviews, while IOM's campaigns attribute phrases, albeit sparingly, to local individuals. This inclusion fosters more humanizing depictions of migration, reframing it not as a mass, depersonalized phenomenon, but as a deeply personal and unique experience. For instance, in *Climate Refugee*, the individual experience of climate-induced migration is poignantly described as 'depressive, difficult, and traumatic.' Similarly, *JCWI's Climate Justice is Migrant Justice* campaign challenges exclusionary policies by centering migrant agencies through the testimonies of Afghan climate activists denied asylum in the UK,. Such narratives resist

[29] https://climatevisuals.org/climate-migration/.

[30] https://climatevisuals.org/climate-migration/.

reductive portrayals of migrants as mere abstractions or statistics, and instead foreground their lived realities.

Yet, as Sinha and Back (2014) caution, the interpretative frames used to understand environmental migration often perpetuate the very problems they set out to address. Many campaigns focus exclusively on case studies from the Global South, inadvertently buttressing racialized imaginaries of environmental migrants as 'the other' (Baldwin, 2022). This framing feeds into what Dal Lago (2012) terms the 'tautology of fear,' whereby displaced individuals are envisioned as outsiders originating from apocalyptic landscapes. For example, *AFD's Climate: All Migrants Tomorrow?* campaign employs technocratic language and visuals to depict migration flows as disruptions to be contained through ecosystem restoration or disaster-response simulations. Such narratives risk perpetuating alarmist registers that obscure systemic inequalities and historical responsibilities bound up with climate injustice. As discussed in Chapter 4, Baldwin (2022) critiques the figure of the 'climate refugee' as a racialized phantom that sustains white supremacy by positioning Northern actors as saviors and erasing their complicity in global ecological harm.

A critical shortcoming of many campaigns lies in their disconnection from broader political or historical contexts. Very few link personal stories to systemic frameworks such as mobility justice or climate justice. An exception is found in the campaigns of *Climate Refugees*. Amali Tower, founder of the NGO, explicitly denounces the lack of international funding from Global North countries—those most responsible for climate change—to support affected nations like Pakistan. This campaign implicitly critiques the Loss and Damage financing mechanism and advocates for structural transformation, rather than merely offering contingent solutions. Similarly, PISFCC's legal advocacy for an ICJ Advisory Opinion integrates Pacific youth voices into global climate governance debates, emphasizing intergenerational equity and demanding accountability from historically polluting states. Sakellari (2024) emphasizes that such approaches are vital for NGOs working in the Global South, especially when advocating for international legal agreements, funding mechanisms, and policies that facilitate human mobility while combating global social injustices. Hence, the most transformative campaigns position climate migrants not only as adaptive agents but also as political subjects demanding justice (Walker & Giacomelli, 2024). By intertwining climate justice with mobility justice, these campaigns challenge depersonalized portrayals of migrants (Fig. 5.1).

Fig. 5.1 Sink/Rise, the third chapter of The Day May Break, Nick Brandt's ongoing global series, highlights those affected by environmental degradation and climate change. This installment centers on South Pacific Islanders facing the threat of rising seas. Captured underwater off the Fijian coast using in-camera techniques, the images show local people who symbolize the many communities at risk of losing their homes, land, and livelihoods as ocean levels continue to rise. Remarkably, though submerged several meters below the surface, the subjects neither float nor swim. Instead, they sit on sofas, stand on chairs, play on seesaws, and pose as if on dry land. This surreal scene suspends the usual laws of physics, creating a haunting space between land and sea. Spending time with these photographs, viewers can sense the shifting emotions reflected in the subjects' faces—stoicism melting into resignation, frustration hardening into resolve—revealing layers of tenderness, grief, and perseverance that quietly ripple beneath the surface and invite deeper contemplation. Courtesy of the artist Nick Brandt

By way of contrast, IOM's campaigns often focus on technical solutions such as constructing seawalls or reinforcing infrastructure without addressing underlying power asymmetries or global inequities. The organization's emphasis on terms like 'internal mobility' and 'forced migration' reflects its alignment with existing legal frameworks but avoids politically charged terminology like 'climate refugees.' While this approach resonates with local contexts—for example, framing migration in Mongolia as historically linked to environmental factors—it risks depoliticizing displacement by elevating technical expertise over systemic critique (IOM Climate Mobility Road Map, 2021–2030). The *Find A Way campaign* acknowledges pre-existing mobility practices but often slips into alarmist tones, as in its depiction of immobility in Madagascar: 'The poorest and eldest have no choice but to stay here and eat the cactus.' Similarly, the *Think of Tomorrow, Act Today* campaign asserts that 'it is vital to accelerate climate action that is underpinned by a comprehensive and holistic understanding of human mobility, which is deeply intertwined with all key dimensions of climate adaptation policy, action and implementation.' Yet, the humanitarian tone of the stories tends to portray people who 'could not adapt to the changing weather' and thus 'have moved away from the community to survive'—or who 'need support and protection to remain in their communities despite their changing environment.'

The contested term 'climate refugee,' used pervasively by *Climate Refugees*, exemplifies the challenges of naming within migration discourse. While intended to foster empathy and advocate for a type of protection akin to refugee status, this term creates a legal non-space for displaced individuals who lack formal recognition under international law (Hiraide, 2022). Moreover, it excludes those (who are the majority) displaced within national borders or immobilized by structural barriers. Baldwin (2022) argues that this label is deeply racialized, evoking anxieties about impending disruption and loss of control in Northern societies. Wisner (2004) further critiques how such terminology fuels xenophobia and racism by framing migrants as threats rather than victims of systemic injustice. From our point of view, it may be more appropriate to adopt the term 'ecologically displaced persons' proposed by Hiraide (2022), which flattens racialized distinctions between North and South while comprising the diverse drivers of displacement beyond climate alone.

Blaming Nature

The representation of climate change in communication campaigns often leans toward framing it as an uncontrollable force of nature, thereby absolving human systems of responsibility. While campaigns such as IOM's *Find A Way, Do the Right Thing,* and *Think of Tomorrow, Act Today* highlight specific stories and contexts, they frequently depict climate change as a natural disaster disconnected from its anthropogenic origins. In the first of these, nature is anthropomorphized as a relentless antagonist. For instance, in Madagascar, drought is described as exacerbated by food insecurity and socioeconomic challenges, while hurricanes in the Bahamas are framed as intensifying due to 'natural disasters.' Similarly, Micronesia is portrayed as a victim of 'Mother Nature's cruelty,' and Allen, a young boy from the Marshall Islands, likens the sea to 'a dog that eats everything it meets.' Similarly, in the video *Think of Tomorrow, Act Today*, the story of the Mbya Guarani community of Santa Teresita reflects the shared experience of many Indigenous communities in South America struggling to protect their ancestral cultures and demonstrating resilience in the face of deforestation and climate change. Although IOM recognizes that 'the forest is at the heart of the Mbya Guaranì culture,' and the protagonist states, 'the forest is a full part of our identity. We have always lived with what nature provides,' there is no mention of the causes, interests, or accountability behind deforestation. These narratives obscure the systemic drivers of vulnerability—colonial histories, extractive economies, and governance failures—and instead cast nature as the sole antagonist (Peters et al., 2019). This metaphor absolves human agency, framing climate change as an uncontrollable force rather than a consequence of systemic inequities.

This apolitical formulation deflects attention from the unequal responsibilities between the Global North and South in contributing to climate change. The International Panel on Climate Change (IPCC) has long emphasized the anthropogenic nature of climate crises, yet IOM's campaigns fail to incorporate this perspective or reference the differentiated responsibilities enshrined in global climate agreements. By portraying injustice as an 'imaginary monster' (Giuliani 2021, p. 140) threatening the Global South, these campaigns perpetuate narratives that divorce human agency from environmental degradation. The visual rhetoric reinforces this abstraction: IOM's *Find A Way* campaign uses filters with white dots (and red for fire in Madagascar) to suggest an ominous force disrupting harmony between people and their environments. Its imagery predominantly depicts post-apocalyptic landscapes, emphasizing adaptation and reconstruction while sidelining mitigation efforts tied to emissions reduction and systemic reform.

This symptomatic strategy calls for a radical reimagining of communication praxis—one that eschews the spectacularization of displacement (Barranquero, 2014) and confronts the colonial logics that reduce the Global South to 'places for disaster' (Giuliani, 2021, p. 141). Such spaces, framed as inherently vulnerable and inhabited by 'non-civilized' others, are paradoxically rendered both hypervisible through humanitarian optics and invisible in terms of historical agency. Visually, IOM employs surreal aesthetics that abstract human–environment connections. Cover photographs feature landscapes filtered with white dots or fiery red hues, symbolizing an ethereal menace. Human figures are often isolated from their surroundings, and stripped of agency. For instance, Jason from the Marshall Islands is shown standing dejectedly by the sea, arms limp by his sides. The accompanying text—'If we don't act now, our islands will disappear'—frames nature as an apocalyptic inevitability while positioning Global North donors as saviors.

In contrast, NGO-led campaigns such as *Climate Justice is Migrant Justice* and Climate Of Change adopt more critical frames that foreground the human-made dimensions of climate crises. JCWI's campaign explicitly links environmental breakdown to exploitative economic systems driven by fossil fuel industries in the Global North. It calls for climate reparations and policy solutions that uphold both the 'right to stay' and the 'right to move,' emphasizing justice over fear-based narratives.[31] Similarly, Climate Of Change frames climate change as a 'man-made crisis,' intertwining ecological destruction with social inequality. This campaign critiques structural inequities while advocating systemic change through youth engagement and participatory storytelling.

The *Climate Refugee* campaign occupies a middle ground. While it provides scientific data on rainfall increases and heatwaves, it occasionally lapses into alarmist tones—'Flood came at 1 a.m. at night. Everything drowned'—without fully addressing root causes like fossil fuel dependency or global economic disparities. However, it also incorporates elements of international climate justice by highlighting Pakistan's limited responsibility for emissions alongside its disproportionate vulnerability to crises exacerbated by Northern policy decisions. In the *Climate*

[31] https://jcwi.org.uk/reportsandbriefings/.

and Migration Coalition campaign,[32] the future is framed in apocalyptic terms, deploying 'narratives of urgency,' which stress the potential scale of climate mobility. These framings, while aimed at catalyzing action, may instead provoke a sense of disorder, panic, or apathy—what Banuledcu-Bogdan et al. (2024) term 'apocalypse fatigue') (Fig. 5.2).

Fig. 5.2 Atlas of the New World. What form might the atlas that guides future generations take? In what ways can a photograph capture what is yet to come? These questions inspired the creation of Atlas of the New World. Traditionally, photography freezes moments in the present, but this project envisions the future as it relates to today's reality. Using a unique slide projector, images depicting potential transformations are cast directly onto existing landscapes. The photographs that emerge are intricate and dreamlike, serving as symbolic glimpses into a near future that awaits us. In the picture: Mozambique is among the most vulnerable African countries to natural disasters. Coastal erosion, frequent droughts, devastating cyclones and floods are episodes that now alternate and recur every year, affecting and destroying entire villages and compromising agricultural harvests.Courtesy of the photographers Giulia Piermartiri and Edoardo Delille

[32] https://climatemigration.org.uk/getting-started-climate-migration/.

Conclusions

As climate change reshapes patterns of migration and displacement, awareness campaigns will continue to play an important role in shaping public understanding of this complex issue. Compelling visual storytelling can effectively engage audiences by conveying detail, nuance, context, and empathy. Yet, communicating climate-linked migration remains a considerable challenge. As we have emphasized throughout this chapter, these information campaigns aim to influence the social imaginary by raising awareness of two of the defining issues of our time. If we consider them as sites where imaginaries are actively constructed, then their narratives must be critically interrogated—not only for what they represent but also for what they erase.

In this chapter, we have seen how awareness campaigns, rooted in colonial legacies, can be understood as contemporary forms of governmentality that operate transnationally, shaping the perceptions of potential migrants from a distance (Heller, 2014; McNevin et al., 2016).

Our analysis sought not only to highlight divergent narrative strategies but also to reveal how these campaigns engage with justice across multiple scales: national, international, intergenerational, and cosmovisional. Moreover, we argued that these campaigns do not merely serve an informational function—they actively shape the 'distribution of the sensible' (Rancière, 2009), determining what is visible, sayable, and thinkable about climate (im)mobility, and whose claims to justice are recognized or denied. In asking what kinds of narrative these campaigns propose about climate (im)mobility, we revealed their deeper purpose: not merely to inform but to border—to inscribe lines that define whose movement is legitimate and whose is suspect; whose suffering is visible and whose remains unseen; whose futures are worth protecting and whose are expendable. In this way, awareness campaigns become architects, not just of discourse, but of bordering futures.

Campaigns led by intergovernmental organizations like the IOM tend to frame climate migration within the bounds of national responsibility and humanitarian sensibility. These often emphasize neoliberal mechanisms like *Loss and Damage*, while largely avoiding a deeper interrogation of the systemic causes of displacement. This approach reinforces a justice paradigm limited to state actors and their national citizens, often neglecting the transboundary and structural dimensions of climate injustice. By contrast, other campaigns—particularly those by international NGOs such as JCWI's *Climate Justice is Migrant Justice* and *Climate Refugee*—foreground the need for cross-border solidarity and legal protections for climate migrants. These initiatives endorse international legal frameworks and policy reforms that acknowledge the global nature of climate-induced displacement along with the shared responsibility of the international community.[33]

Some campaigns, notably the *Pacific Islands Students Fighting Climate Change* (PISFCC), explicitly invoke intergenerational equity, emphasizing the moral and legal obligations of the present generation to safeguard the rights and well-being of future generations. This orientation is reflected in their advocacy for legal precedents and policy solutions that account for the long-term impacts of climate change, including youth-led climate litigation and international legal initiatives.

A few campaigns attempt to incorporate cosmovisional perspectives, centering the worldviews and knowledge systems of Indigenous and marginalized communities. By amplifying place-based narratives and connecting environmental degradation to colonial histories and land dispossession, these campaigns challenge dominant epistemologies and issue a call for justice that honors diverse relationships with land, water, and non-human life.

[33] https://climatejustice.uk/migrantjustice/.

Campaign	*National justice*	*International justice*	*Intergenerational justice*	*Cosmovisional justice*	*Quotes from campaign text*
Do the Right Thing (IOM)	Focuses on national-level humanitarian response, loss and damage, but does not critique deeper systemic or global causes	Limited	Not explicit	Not explicit	IOM's campaigns rely on humanitarian imagery, portraying migrants as victims in need of salvation

(continued)

(continued)

Campaign	*National justice*	*International justice*	*Intergenerational justice*	*Cosmovisional justice*	*Quotes from campaign text*
Climate Refugees (Climate Refugees)	Documents national and local impacts	Advocates for international legal protection and policy change	Highlights future risks and rights	Centers Global South voices, challenges Eurocentric narratives	'We at Climate Refugees believe the time has come to act now to ensure the international community stands ready to protect those who are forced to flee their homes or even their countries because of climate change… We document cases to shed light on human rights impacts, protection gaps, possible legal voids and advocate human rights-based solutions for the creation of policies and legal norms that protect those impacted by all effects of climate change-induced displacement and migration.'[34]

(continued)

[34] https://www.climate-refugees.org/foundingvision.

(continued)

Campaign	*National justice*	*International justice*	*Intergenerational justice*	*Cosmovisional justice*	*Quotes from campaign text*
Together, We Can #FindAWay (IOM)	Focuses on national-level humanitarian response, loss and damage, but does not critique deeper systemic or global causes	Limited	Not explicit	Not explicit	Not directly cited, but IOM campaigns generally lean heavily on victimhood tropes while tending to obscure systemic inequalities
Think of Tomorrow, Act Today (IOM)	Focuses on international level and humanitarian response, but does not critique deeper systemic or global causes	Advocates for international solidarity and policy change	Explicitly includes youth voices	Centers voices of affected communities	'The time to act is now.' 'Each year, he watches the land shrink but remains determined to keep it alive for future generations.' 'Be a lifeline to people in need. With your help, IOM can save lives by providing crucial assistance wherever and whenever it's needed the most'[35]

(continued)

35 https://acttoday.iom.int/#:~:text=At%2017%2C%20Allen%20feels%20an,families%20and%20reshaping%20entire%20communities.

(continued)

Campaign	*National justice*	*International justice*	*Intergenerational justice*	*Cosmovisional justice*	*Quotes from campaign text*
Climate Justice is Migrant Justice (JCWI)	Addresses UK national policy and legal context	Advocates for international solidarity and legal precedents	Explicitly includes youth voices	Centers voices of those forced to move, challenges colonial and racialized narratives	'We believe that, in the face of these challenges, everyone should have the right to stay and the right to move… Our legal team will represent people who've had to cross borders because of climate breakdown, to protect their individual rights and establish groundbreaking legal precedents for the protection of other people who move… We'll be establishing a steering committee of people who've been forced to move because of the impacts of the climate crisis.'[36]

(continued)

[36] https://jcwi.org.uk/campaigns/climate-justice-is-migrant-justice/.

(continued)

Campaign	*National justice*	*International justice*	*Intergenerational justice*	*Cosmovisional justice*	*Quotes from campaign text*
Climate: All Migrants Tomorrow? (AFD Group)	Focus on national and local adaptation policies	Supports interna-tional cooperation and funding	Not explicit	Not explicit	'AFD Group is deploying a strategy based on three axes. First, aim to prevent migration induced by climate change, by supporting public policies for mitigating and adapting to climate change, by financing projects to restore ecosystems and maintain natural resources, or reducing people's exposure to disasters.'[37]

(continued)

[37] https://www.afd.fr/en/climate-all-migrants-tomorrow.

(continued)

Campaign	*National justice*	*International justice*	*Intergenerational justice*	*Cosmovisional justice*	*Quotes from campaign text*
(Climate and Migration Coalition)	Advocates for national and international policy change	Focus on international protection and rights	Not explicit	Centers voices of affected communities	'Our goal is to ensure a people centred policy response at the national and international level… Much of our work focuses on the testimonies and stories of people who are already moving due to climate change impacts. We believe that these voices should be central to the debate about how this issue is addressed.'[38]

(continued)

[38] https://weadapt.org/organisation/uk-climate-change-and-migration-coalition/.

(continued)

Campaign	*National justice*	*International justice*	*Intergenerational justice*	*Cosmovisional justice*	*Quotes from campaign text*
Pacific Islands Students Fighting Climate Change (PISFCC)	Engages national leaders in Pacific states	Seeks international legal recognition (ICJ advisory opinion)	Explicitly foregrounds rights of current and future generations	Draws on Pacific worldviews and youth activism	'Seek an advisory opinion on the obligations of states, under international law, towards current and future generations in protecting their rights from the adverse impacts of climate change… Educate and activate Pacific youth to become more conscious and aware of the single greatest threat to their homes, lives and their future.'[39]

Source Authors' elaboration.

[39] https://www.pisfcc.org/news/student-activists-take-global-climate-fight-forward.

The campaigns' divergent paths reveal a contested terrain where communication is both a weapon and a wound. To transcend the Panicocene, campaigns must embrace what Gidley et al. (2009) names 'participatory futuring': the co-creation of stories with migrants as co-authors, not just subjects or mere objects of representation. This demands dismantling the 'observant eye' of colonial visuality (Knudsen et al., 2022) and adopting what Alqaisiya (2023) calls place-based tale-telling—narratives in which Haitian farmers articulate land theft alongside drought, and Tunisian youth link EU border deals to desertification. Only then can climate (im)mobilities be reimagined not as apocalyptic specters, but as crucibles of global repair and healing.

In the shimmering landscape of the mediapolis, the stories we tell about climate (im)mobility are neither neutral nor incidental. They are battlegrounds of meaning, where awareness campaigns wield language both as a compass and a cage—illuminating crises while casting shadows of exclusion. These campaigns, circulating through intergovernmental reports and viral hashtags, do more than inform: they inscribe moral cartographies, drawing symbolic borders that prefigure, legitimize, and complement physical ones (Musarò, 2019). As algorithms and infographics dissect the entanglement of melting glaciers and human movement, they resurrect age-old dichotomies—care and control, threat and victimhood—that haunt the geopolitics of mobility.

The mediapolis, now a hybrid ecosystem of TikTok testimonials and UN white papers, no longer merely reflects reality but actively assembles it. Whereas Buraschi et al. (2017) observed its power to elevate or erase issues, we now confront it as a double-edged instrument: campaigns like *Do the Right Thing* (IOM) and *Think of Tomorrow, Act Today* deploy empathy-inducing visuals of submerged villages, yet often reduce complex climate mobilities to a binary of 'worthy' displacement or 'risky' invasion. Foucault's (1971) 'order of discourse' manifests here not just in headlines but in the pixelated gaze of drone footage—a gaze that, as Musarò (2019) notes, wavers between humanitarian pathos and securitarian panic.

To speak of climate migrants in the mediapolis is to navigate a semantic minefield. Terms like 'climate refugee' float in a legal limbo yet get crystallized in the public imagination as racialized specters (Baldwin & Bettini, 2017). Campaigns may seek to foster solidarity but often risk reducing individuals to 'tragic icons' (Chouliaraki, 2013)—a depoliticized shorthand that obscures colonial legacies and neoliberal extractivism. Herein lies the paradox: awareness-raising, framed as a tool for justice, becomes a symbolic checkpoint, filtering narratives through the barbed wire of 'managed' compassion.

The Panicocene emerges not only from climate chaos but from the stories that make it intelligible. When *Climate Refugees* centers voices from the Global South, it disrupts the Eurocentric dramaturgy of rescue—yet such efforts remain glaring exceptions. A decolonial lens reveals how many campaigns replicate colonial epistemologies by centering Global North perspectives and universalizing vulnerability while silencing marginalized voices. They reproduce what Fairclough (2001) calls 'technologies of consent,' embedding hierarchies through lexical choices (e.g., 'flows' vs. 'exoduses') and visual tropes (sun-bleached mothers clutching children vs. 'resilient' entrepreneurs). These narratives, as intersectional critique shows, often erase gendered and racialized vulnerabilities, naturalizing displacement rather than interrogating the systems that produce it.

This chapter has traced these tensions through campaigns that straddle pedagogy and propaganda, arguing that every infographic is a micro-border—a site where worlds are both imagined and policed. The 'soft power' of awareness campaigns lies in their capacity to 'prefigure the hard borders of tomorrow' (Musarò, 2019,) making the rewriting of their scripts not just an academic exercise but an act of radical care. Each campaign employs a distinct blend of textual narratives and visual rhetoric to shape public understanding. Visual elements found across campaigns—whether through surreal filters that abstract human–environment relations or intimate interviews that amplify migrant voices—underscore how representation shapes public perception and, ultimately, policy. IOM's campaigns lean heavily on humanitarian imagery, portraying migrants as passive victims in need of salvation. This reliance on victimhood tropes obscures the systemic inequalities and anthropogenic causes of climate change. By contrast, *JCWI* adopts a progressive framing that emphasizes resilience, agency, and justice—recasting migration as a matter of rights and solidarity rather than security. The *Climate Refugee* campaign foregrounds systemic inequalities and advocates for climate justice. *AFD*, in turn, advances a technocratic narrative, emphasizing prevention and adaptation through ecosystem restoration and institutional reform. By unraveling the framing architectures of these campaigns, we uncover how public perception and policy responses are shaped. These campaigns distill the chaotic entanglement of climate collapse and human mobility into archetypes: the 'climate refugee' as tragic victim, and the 'adaptive migrant' as a good neoliberal subject. Though legally nebulous, these figures crystallize in the collective psyche, racializing mobility (Baldwin & Bettini, 2017) and depoliticizing displacement. The visual grammar of these campaigns—a child knee-deep in floodwater, a parched terrain enclosed by fences—does not simply document suffering; it aestheticizes it, rendering structural violence a natural disaster. As Bartoletti and Faccioli (2013) note, the mediapolis thrives on polyphony: grassroots collectives hack dominant frames and algorithms to recenter marginalized epistemologies. Yet a decolonial lens reveals that this heteroglossia is profoundly asymmetric. When campaigns reduce climate (im)mobility to technical 'solutions' or apolitical 'adaptation,' they erase the colonial histories and capitalist extractivism that fracture ecosystems and displace communities.

Ultimately this chapter has argued for the construction of a counter-cartography—one that resists pathologization and racialization by amplifying justice-oriented narratives over frameworks of charity, pity, or control. In our view, these campaigns are far from mere conduits of information: they are active players in shaping 'moral geographies' that influence public consciousness. They determine who is granted visibility and whose existence fades into obscurity; which narratives are amplified, and which are confined to silence. In doing so, they inscribe symbolic boundaries—borders that parse out inclusion from exclusion, compassion from apathy, and solidarity from indifference.

To reimagine climate (im)mobility is to embrace its poetic plurality—a tapestry woven from resilience, struggle, agency, and interconnectedness across global divides. By moving beyond alarmist tropes toward inclusive narratives rooted in justice, mutual aid, accountability, and reparative action, we can begin to trace new cartographies of solidarity that honor both human dignity and planetary interdependence.

References

Alqaisiya, W. (2023). The decolonial wor(l)ds of Indigenous women. *Social & Cultural Geography, 25*(7), 1025–1043. https://doi.org/10.1080/14649365.2023.2268583

Amnesty International. (1997). *Amnesty International Campaigning Manual*. Amnesty International; International Service for Human Rights.

Anne DiFrancesco, D., & Young, N. (2011). Seeing climate change: The visual construction of global warming in Canadian national print media.

Cultural Geographies, 18(4), 517–536. https://doi.org/10.1177/1474474010382072

Baldwin, A. (2022). *The other of climate change: Racial futurism, migration, humanism*. Rowman & Littlefield.

Baldwin, A., & Bettini, G. (Eds.). (2017). *Life adrift: Climate change, migration, critique*. Rowman & Littlefield International.

Banulescu-Bogdan, N., & Huang, L. (2024). *Public opinion of climate migrants: Understanding what factors trigger anxiety or support*. Migration Policy Institute.

Barranquero, A. (2014). Comunicación, cambio social y ONG en España. Pistas para profundizar en la cultura de la cooperación desde los nuevos movimientos comunicacionales. El caso del 15M. *COMMONS. Revista de Comunicación y Ciudadanís Digital, 3*(1), 6–28.

Bartoletti, R., & Faccioli, F. (2013). *Comunicazione e civic engagement: Media, spazi pubblici e nuovi processi di partecipazione*. F. Angeli.

Bishop, S. C. (2020). An international analysis of governmental media campaigns to deter asylum seekers. *International Journal of Communication, 14*, 1092–1114.

Boas, I. (2015). *Climate migration and security*. Routledge. https://doi.org/10.4324/9781315749228

Boas, I., Farbotko, C., Adams, H., Sterly, H., Bush, S., van der Geest, K., Wiegel, H., Ashraf, H., Baldwin, A., Bettini, G., Blondin, S., de Bruijn, M., Durand-Delacre, D., Fröhlich, C., Gioli, G., Guaita, L., Hut, E., Jarawura, F. X., Lamers, M., & Hulme, M. (2019). Climate migration myths. *Nature Climate Change, 9*(12), 901–903. https://doi.org/10.1038/s41558-019-0633-3

Boas, I., Wiegel, H., Farbotko, C., Warner, J., & Sheller, M. (2022). Climate mobilities: Migration, im/mobilities and mobility regimes in a changing climate. *Journal of Ethnic and Migration Studies, 48*(14), 3365–3379. https://doi.org/10.1080/1369183X.2022.2066264

Brambilla, C. (2021). Revisiting 'bordering, ordering and othering': An invitation to 'migrate' towards a politics of hope. *Tijdschrift Voor Economische En Sociale Geografie, 112*(1), 11–17. https://doi.org/10.1111/tesg.12424

Brändle, V. K. (2022). Well informed? EU governments' digital information campaigns for (potential) migrants. *Journal of Immigrant & Refugee Studies, 22*(3), 467–481. https://doi.org/10.1080/15562948.2022.2046896

Browne, E. (2015). *Impact of communication campaigns to deter irregular migration* (1248; GSDRC Helpdesk Research Report).

Buraschi, D., & Aguilar Idáñez, M. J. (2016). Indiferencia, fronteras morales y estrategias de resistencia. *Documentación Social, 180*, 127–147.

Buraschi, D., Aguilar Idáñez, M. J., Oldano, N., Fonte García, M. E., & Zapata Hernández, V. M. (2017). Comunicazione partecipativa e cultura dell'ospitalità. *Educazione Aperta, Rivista Di Pedagogía Critica, 2*, 11–32.

Butler, J. (2015). *Notes toward a performative theory of assembly*. Harvard University Press.
Cappi, V., & Musarò, P. (2022). Awareness campaigns to deter migrants: A neoliberal industry for symbolic bordering. In A. Dastyari, A. Nethery, & A. Hirsch (Eds.), *Refugee externalisation policies: Responsibility, legitimacy and accountability* (pp. 171–188). Routledge.
Chouliaraki, L. (2013). *The ironic spectator: Solidarity in the age of post-humanitarianism*. Polity Press.
Chouliaraki, L. (2024). *Wronged: The weaponization of victimhood*. Columbia University Press.
Clarke, J., Webster, R., & Corner, A. (2020). *Theory of Change: Creating a social mandate for climate action*. Climate Outreach.
Clement, V., Rigaud, K. K., de Sherbinin, A., Jones, B., Adamo, S., Schewe, J., Sadiq, N., & Shabahat, E. (2021). *Groundswell Part 2: Acting on Internal Climate Migration*. World Bank. https://openknowledge.worldbank.org/handle/10986/36248
Climate Outreach. (2024). *Climate and migration: A user's guide to navigating the research*. https://Climateoutreach.Org/Climate-and-Migration-a-Users-Guide-to-Navigating-the-Research/
Dal Lago, A. (2012). *Non-persone: L'esclusione dei migranti in una società globale* (Nuova ed., 5. ed). Feltrinelli.
De Haas, H. (2020). *Climate refugees: The fabrication of a migration threat*. http://Heindehaas.Blogspot.Com/2020/01/Climate-Refugees-Fabrication-of.html
Dupar, M. K., McNamara, L., & Pacha, M. (2019). *Communicating climate change: A practitioner's guide*. Climate and Development Knowledge Network.
Dziminska, M., Mendoza, I., Pellegrini, G., & Rowland, J. (2021). Climate change and public perception. Citizens' proposals for better communication and involvement. *Journal of Science Communication, 20*(03), A09. https://doi.org/10.22323/2.20030209
Fairclough, N. (2001). *Language and power* (2nd ed). Longman.
Fiedler, A. (2020). From being aware to going there: On the awareness and decision-making of (prospective) migrants. *Mass Communication and Society, 23*(3), 356–377. https://doi.org/10.1080/15205436.2019.1666992
Foucault, M. (1971). Orders of discourse. *Social Science Information, 10*(2), 7–30. https://doi.org/10.1177/053901847101000201
Foucault, M. (1980). *Power/knowledge: Selected interviews and other writings, 1972–1977* (1st American ed). Pantheon Books.
Giacomelli, E., & Cappi, V. (2025). 'Othering' in the Panicocene: Intergovernmental and non-governmental awareness campaigns on climate change

induced-migration. *Journal of Ethnic and Migration Studies, 1–19*. https://doi.org/10.1080/1369183X.2025.2468287

Giacomelli, E., & Walker, S. (2022). Contronarrazioni del cambiamento climatico. Diari di ingiustizia climatica tra Dakar e Saint Louis, Senegal. *Sociologia della Comunicazione, 64*, 121–140. https://doi.org/10.3280/SC2022-064007

Giacomelli, E., & Walker, S. (2025). Climate diaries. In L. Oso, N. Ribas-Mateos, & M. Moralli (Eds.), *Elgar Encyclopedia of global migration* (pp. 123–125). Edward Elgar Publishing. https://doi.org/10.4337/9781035300389.ch33

Gidley, J. M., Fien, J., Smith, J., Thomsen, D. C., & Smith, T. F. (2009). Participatory futures methods: Towards adaptability and resilience in climate-vulnerable communities. *Environmental Policy and Governance, 19*(6), 427–440. https://doi.org/10.1002/eet.524

Giuliani, G. (2021). *Monsters, catastrophes and the anthropocene: A postcolonial critique*. Routledge.

Hartmann, B. (2010). Rethinking climate refugees and climate conflict: Rhetoric, reality and the politics of policy discourse. *Journal of International Development, 22*(2), 233–246. https://doi.org/10.1002/jid.1676

Heller, C. (2014). Perception management – Deterring potential migrants through information campaigns. *Global Media and Communication, 10*(3), 303–318. https://doi.org/10.1177/1742766514552355

Hiraide, L. A. (2022). The difficult business of defining climate refugees. *Green European Journal*. https://www.greeneuropeanjournal.eu/the-difficult-business-of-defining-climate-refugees/

Knudsen, B. T., Oldfield, J. R., Buettner, E., & Zabunyan, E. (Eds.). (2022). *Decolonizing colonial heritage: New agendas, actors and practices in and beyond Europe*. Routledge.

Kosnick, K. (2014). Mediating migration: New roles for (mass) media. *InMedia, 5*. https://doi.org/10.4000/inmedia.761

Koteyko, N., & Atanasova, D. (2016). Discourse analysis in climate change communication. In N. Koteyko & D. Atanasova (Eds.), *Oxford Research Encyclopedia of Climate Science*. Oxford University Press. https://doi.org/10.1093/acrefore/9780190228620.013.489

Kress, G. R., & Leeuwen, T. van. (2010). *Reading images: The grammar of visual design* (2. ed., reprinted). Routledge.

Manuti, A. (2013). Climate change awareness: An explorative study on the discursive construction of ethical consumption in a communication campaign. *American Journal of Applied Psychology, 1*, 65–71.

McNevin, A., Missbach, A., & Mulyana, D. (2016). The rationalities of migration management: Control and subversion in an Indonesia-based counter-smuggling campaign. *International Political Sociology, 10*(3), 223–240. https://doi.org/10.1093/ips/olw009

Methmann, C., & Oels, A. (2015). From 'fearing' to 'empowering' climate refugees: Governing climate-induced migration in the name of resilience. *Security Dialogue, 46*(1), 51–68. https://doi.org/10.1177/0967010614552548

Movement Advancement Project and GLAAD. (2008). *Communications campaign best practices*. https://www.lgbtmap.org/communications-campaign-best-practices

Musarò, P. (2019). *Aware Migrants*: The role of information campaigns in the management of migration. *European Journal of Communication, 34*(6), 629–640. https://doi.org/10.1177/0267323119886164

Nieuwenhuys, C., & Pécoud, A. (2007). Human trafficking, information campaigns, and strategies of migration control. *American Behavioral Scientist, 50*(12), 1674–1695. https://doi.org/10.1177/0002764207302474

Nixon, R. (2013). *Slow violence and the environmentalism of the poor* (First Harvard University Press paperback edition). Harvard University Press.

Olausson, U. (2009). Global warming—global responsibility? Media frames of collective action and scientific certainty. *Public Understanding of Science, 18*(4), 421–436. https://doi.org/10.1177/0963662507081242

Organizzazione Internazionale per le Migrazioni (OIM). (2020). *IOM Public Communication Campaign Toolkit*. Internat. Organization for Migration.

Pécoud, A. (2010). Informing migrants to manage migration? An analysis of IOM's information campaigns. In M. Geiger & A. Pécoud (Eds.), *The politics of international migration management* (pp. 184–201). Palgrave Macmillan. https://doi.org/10.1057/9780230294882_9

Peters, K., Mayhew, L., Slim, H., van Aalst, M., & Arrighi, J. (2019). *Double vulnerability: The humanitarian implications of intersecting climate and conflict risk*. ODI. https://Odi.Org/En/Publications/Double-Vulnerability-the-Humanitarian-Implications-of-Intersecting-Climate-and-Conflict-Risk/

Raimi, K. T., Sarge, M. A., Geiger, N., Gillis, A., & Cunningham, J. L. (2024). Effects of communicating the rise of climate migration on public perceptions of climate change and migration. *Journal of Environmental Psychology, 93*, Article 102210. https://doi.org/10.1016/j.jenvp.2023.102210

Ransan-Cooper, H., Farbotko, C., McNamara, K. E., Thornton, F., & Chevalier, E. (2015). Being(s) framed: The means and ends of framing environmental migrants. *Global Environmental Change, 35*, 106–115. https://doi.org/10.1016/j.gloenvcha.2015.07.013

Rancière, J. (2009). *The politics of aesthetics: The distribution of the sensible* (Reprinted). Continuum.

Rice, R. E., & Atkin, C. K. (2013). Theory and principles of public communication campaigns. In R. E. Rice & C. K. Atkin (Eds.), *Public communication campaigns* (4th ed, pp. 16–50). Sage.

Rose, G. (2022). *Visual methodologies: An introduction to researching with visual materials* (Fifth). Sage.

Sakellari, M. (2024). Communicating climate change induced migration: The role of NGOs. *Open Research Europe, 3*, 163. https://doi.org/10.12688/openreseurope.16232.2

Schans, D., & Optekamp, C. (2016). *Raising awareness, changing behavior? Combating Irregular Migration through Information Campaigns*. Ministry of Justice and Security. https://www.eerstekamer.nl/overig/20170328/raising_awareness_changing

Schloenhardt, A., & Philipson, E. (2013). No to people smuggling': A review of Australia's antimigrant smuggling awareness campaigns. *The University of Queensland Migrant Smuggling*. https://law.uq.edu.au/files/6729/Schloenhardt-Philipson-SoM-Awareness-Campaigns-2013.pdf

Silverstone, R. (2009). *Mediapolis: La responsabilità dei media nella civiltà globale*. V & P.

Sinha, S., & Back, L. (2014). Making methods sociable: Dialogue, ethics and authorship in qualitative research. *Qualitative Research, 14*(4), 473–487. https://doi.org/10.1177/1468794113490717

Tilley, L., Ranawana, A. M., Baldwin, A., & Tully, T. M. (2023). Race and climate change: Towards anti-racist ecologies. *Politics, 43*(2), 141–152. https://doi.org/10.1177/02633957221127166

Tjaden, J., Morgenstern, S., & Laczko, F. (2018). *Evaluating the impact of information campaigns in the field of migration: A systematic review of the evidence and practical guidance* (Central Mediterranean Route Thematic Report Series). International Organization for Migration.

Trauner, F., Brekke, J.-P., Adam, I., Cham, O. N., & Thorbjørnsrud, K. (2024). 'It's not about the information, it's about the situation': Understanding the misalignment between EU deterrence messaging and migrants' narratives. *Journal of Ethnic and Migration Studies, 50*(14), 3379–3395. https://doi.org/10.1080/1369183X.2024.2333865

Udah, H. (2024). Decolonising research for justice: Ethical imperatives and practical applications. *International Journal of Qualitative Methods, 23*, 16094069241294040. https://doi.org/10.1177/16094069241294040

van Dijk, T. A. (1994). Critical discourse analysis. *Discourse & Society, 5*(4), 435–436. https://doi.org/10.1177/0957926594005004001

Walker, S., & Giacomelli, E. (2024). Encountering mobility (in)justice through the lived experiences of fishing communities in Dakar and Saint Louis, Senegal. *Mobilities, 19*(6), 955–971. https://doi.org/10.1080/17450101.2024.2334705

Weiss, J. A., & Tschirhart, M. (1994). Public information campaigns as policy instruments. *Journal of Policy Analysis and Management, 13*(1), 82. https://doi.org/10.2307/3325092

Wisner, B. (Ed.). (2004). *At risk: Natural hazards, people's vulnerability, and disasters* (2nd ed). Routledge.

CHAPTER 6

Creative Perspectives on Climate (Im)Mobilities

Introduction

What is worth more, art or life? During the German environmental group Last Generation's attack on Van Gogh's *Sunflowers* in London, *Just Stop Oil* protesters shouted this question. Members of *Last Generation*—who had also been behind the attack on the Monet painting in the Potsdam Museum—insisted on the sentiment in a post following their acts of dissent, asking the same question through their virtual megaphone. In recent years, there (Chouliaraki 2024) have been several attacks on works of art in international museum collections. Demanding urgent climate action, activists from various activist groups have thrown soup and glued themselves to famous artworks around the world. Recent climate activism has included a series of high-profile interventions in art institutions, such as the splashing of pea soup on a Van Gogh painting in Rome, the campaigners of Extinction Rebellion targeting a Picasso work in Melbourne, and activists affixing themselves to artworks by Botticelli, Boccioni, Van Gogh, and other renowned artists. Crucially, none of the targeted works sustained permanent damage, as most of them are protected by glass. These actions appear strategically designed to draw media attention to the escalating consequences of the climate crisis, rather than to inflict harm on the artworks themselves. In response, nearly one hundred directors of major cultural institutions have signed a public statement condemning these acts, arguing that museums should serve as

E. Giacomelli and P. Musarò, *Climate Mobility Justice*, https://doi.org/10.1007/978-3-032-07907-7_6

spaces for 'social discourse'[1] rather than as sites for protest actions that risk damaging cultural heritage.

Concurrently, contemporary art has seen a resurgence of creative practices that not only critique the destructive logic of our production system but also draw meaning from the performative rituals of resistance. The *Art for Radical Ecologies Manifesto* (Baravalle et al., 2024)–which originated from the *Venice Climate Camps* (2022) and debuted at the *World Congress for Climate Justice* in Milan (2023)—expresses 'a strong rejection of the function of art in the commodification of contemporary techno-financial turbo-capitalism: but at the same time, an attempt to redefine the territory of art in its capacity to imagine other worlds and, at the same time, be the finite subject that materially generates its perception on the brink of the conflicting becoming of time' (Braga, 2024, p. 38). Emanuele Braga is one of the founders of this international community of artists engaged within the context of radical eco-activism. In a statement contesting art that stops at speculation—that is, which avoids awareness of what it generates in terms of real, bodily, and material relationships—he writes that 'the role of art is to occupy the imaginary by de-automating the machine that is building our bodies' (2024, p. 41). Organizing marches, managing clashes with the police, trespassing on construction sites and private properties, and sharing communal life with activists from around the world, the collective *Art for Radical Ecologies* (Baravalle et al., 2024) opens up spaces that resist colonial privilege and complicity with biocapitalism, extractive industry, and financial greenwashing while also taking a strong stance against the containment and detention of migrating humans and other-than-humans. Although not explicitly stated in their *Manifesto*, these 'artivists' use their art to foster ecocentrism and interspecies mutualism—an art of the planetary commons—reclaiming climate and mobility justice by disrupting colonial, anthropocentric, and capitalist narratives.

In this chapter, we explore art as a means through which we see, feel, and reimagine the world. To ground our critical analysis of dominant approaches, discourses, and tools currently used to address climate and mobility justice, we focus on the role of art in creating new narratives and

[1] The statement was released by the International Council of Museums in Germany. Cfr: https://icom-deutschland.de/de/nachrichten/564-statement-attacks-on-artworks-in-museums.html.

challenging dominant imaginaries. Artistic practices can cultivate inclusive and collaborative spaces of expression that, on one hand, deconstruct dominant rhetoric of migrants as passive victims or potential threats, and on the other, amplify and synchronize diverse voices in public discourse. In doing so, they add affective substance and aesthetic power to the instances and examples they convey (Moralli et al., 2021). Artistic creation thereby embodies a form of 'politicality': a space of conviviality and positive hybridity, where people interact and perform both commonality and difference, perceiving themselves as part of the same symbolic, cognitive, and physical universe without erasing diversity (Paltrinieri et al., 2020).

But what role does art play in the age of Panicocene (Giacomelli, 2023)? Artistic practices and artworks are increasingly recognized as essential tools for the dismantling of dominant narratives and the opening of alternative spaces of imagination, representation, and action (Musarò et al., 2022). In the face of climate change-induced (im)mobilities, artistic interventions can serve as compelling counterpoints to hegemonic discourses that frame migrants either as burdens or menaces rather than as agents of history and subjects of injustices. These practices have the power to stimulate new imaginaries by cultivating a politics of care, reciprocity, and responsibility toward otherness and planetary life.

In the following pages, we explore how contemporary art navigates the tension between avant-garde disruption and decolonial care, destabilizing the binary between shock and stewardship. While the avant-garde tradition embraces rupture in order to jolt audiences into climate awareness, decolonial approaches—rooted in Indigenous cosmologies and relational accountability—tend to emphasize healing and interconnection. Though seemingly at odds, these two modes form a vital polyphony, with each counterbalancing the specters of institutional co-optation; namely the risks of seeing avant-garde commodified as cultural capital, and that decolonial aesthetics be reduced to superficial diversity and inclusion. In the face of the Panicocene, both strategies are vital. Shock and urgency must dismantle extractive systems; care must envision alternative futures grounded in ancestral wisdom. Together, shock and care emerge not as opposites but as interdependent responses to ecological crisis and emotional paralysis.

Art in relation to climate (im)mobilities manifests in various forms—from 'eco-arts,' which makes visible the imperceptible effects of humans on flora and fauna (Cucuzzella, 2021), to 'artivism,' which forges links between direct action, political claims, and artistic tactics, and socially engaged art rooted in community participation. Artistic and creative methods of audience engagement effectively overcome participation barriers, stimulating collective confidence and self-expression (Seiger et al. 2022).

At the same time, the work of scholars like Yumagulova et al. (2023) offers a powerful lens for understanding how these narratives intersect with Indigenous perspectives on climate mobility justice. Their research critiques dominant climate displacement frameworks that erase Indigenous agency and governance, advocating instead for self-determined adaptation strategies. These insights are essential for analyzing artistic responses to climate (im)mobility, especially those that emerge from or engage with Indigenous communities.

How are contemporary artistic practices engaging with climate change-induced (im)mobilities? What is the role of art in challenging the hegemonic gaze of the Global North? How can we recognize migration not as an exception but as a constitutive feature of our world—a lens for experiencing existence and grasping our being in the world? How might performing arts offer new insights into the entanglements of climate and mobility justice?

We pursue these questions by exploring how some contemporary artistic practices have compellingly reshaped representations of climate (im)mobilities. More specifically, we look at the work of Angela Tiatia, Andrea Conte (Andreco), Jason deCaires Taylor, John Akomfrah, Ropate Kama, the collective Mali Weil, Yinka Shonibare, Lena Dobrowolska, and Teo Ormond-Skeaping, as we track the changes that art brings into the world (Meskimmon, 2023). Following Terry Smith's lead, we will show how by deploying global media tools, participating in transnational activist networks, and initiating new civic platforms for cross-cultural exchanges, artists 'imagine the world as a differentiated yet inevitably connected whole' (Smith, 2011, p. 8).

We have focused on the above-mentioned artists because their work enacts mainly non-Western cultural practices tackling the complexities of ecological and cultural transformation, while also displaying a critical consciousness of (im)mobilities in relation to the legacies of decolonialism and other (in)justices. By focusing on the aesthetics and politics of these works, this chapter examines how artists question, destabilize, and reconstruct narratives surrounding climate displacement, adaptation, and justice. The aim is to reveal art's capacity to challenge that relationship, inventing a set of interconnected calls for more mutual forms of coexistence: calls to historicize, to become responsible, to empathize, to story-tell. In so doing, arts contribute by suggesting new meanings rather than imposing them, questioning rather than dispensing answers, and empowering the spectator to do his or her own 'borderwork' (Rumford, 2014) or 'de-borderwork' between self and other in a changing world (Khosravi, 2011).

The Ethics of Showing and the Ethics of Seeing

Examining a diverse array of artistic interventions into the discursive regime that frames the Mediterranean border spectacle of desperation and mass death, in the book *Reframing Migration* (2019) Federica Mazzara explores how activists and art forms have become platforms for subverting the dominant narrative of migration and generating new forms of political dissent. Similarly, in her book *Art for Coexistence* (Ross, 2022), art historian Christine Ross examines contemporary art's response to migration, showing that art invites us to abandon our preconceptions about the current 'crisis' and to see migration more disobediently. Revisiting migration experiences through the lens of art, she explores the practices of several artists to demonstrate art's power to mediate experiences of migration. By radically critiquing the legacy of imperialist museology—which dissocializes, dehistoricizes, and depoliticizes art—Ross argues that art invents a set of interconnected calls for more mutual forms of coexistence: to unlearn colonization, to become responsible, to empathize, and to tell stories.

Mazzara and Ross, among others, stress the role of art in inspiring empathy, understood both as an affective response to another person's emotional state and as a cognitive capacity to understand their perspective. They explore empathy's potential in artistic practices motivated by a desire to respond prosocially to the injustices, inequalities, exclusions, and distress that shape twenty-first-century migration. They emphasize the centrality of empathy in the making and reception of artworks and images, and in what cultural studies scholar E. Ann Kaplan (2011, p. 51), in her study of empathy as a possible form of witnessing, describes as 'feeling so shocked by suffering that one is moved to act… . [O]ne is motivated to see that justice is done.'

While following this approach, we remain fully aware of empathy's ambivalence, and of the fact that, on its own, empathy is simply not enough. As several studies (Singer, 2011; Bloom, 2018) argue, empathy may be biased toward members of one's group, failing to extend beyond those who are socially, culturally, and politically similar to the empathizer. This limitation speaks to the narrow scope and moral shortcomings of empathy. As Breithaupt argues in *The Dark Sides of Empathy* (Breithaup & Hamilton 2019), empathy can reinforce social boundaries, stigma, and discrimination. More broadly, empathy studies urge vigilance in our empathic responses, especially when reacting to others' suffering—a warning that is especially important to heed in contexts of migration or climate change.

As sociologists and media scholars, we are aware that the proliferation of images and stories about vulnerable others has led to widespread pessimism. Several late twentieth-century photography critics (Barthes, 1982; Sontag, 2013) argued that representations of victimization offer only a grim choice between narcissistic identification and voyeurism. Images and stories can desensitize us to suffering, undermine their moral urgency, and contribute to apathy among media publics—a condition Susan Moeller (Moeller 1999) has termed 'compassion fatigue.' As Chouliaraki (2006, 2013) argues, this skepticism toward mediation fosters a dystopian view of solidarity: the media commodifies distant suffering, transforming other-oriented dispositions into cynical hyperindividualism.

We understand that both the medium and the message deserve close scrutiny, particularly regarding their effects. For instance, photography bears a witnessing power that can serve multiple purposes: while it may soberly document the destruction of soil and landscape by human industry, it may also arouse a visceral, shocking reaction in viewers. Shock, however, as Susan Sontag (2013) famously warns, 'can become familiar. Shock can become exhausted.' The imagery meant to awaken us to environmental devastation and disaster has become so ubiquitous that it often leads to desensitization. As visual culture theorist Nicholas Mirzoeff observes, 'the Anthropocene is so embedded in our senses that it determines our perceptions, therefore it is aesthetic' (Mirzoeff, 2014, p. 223).

At the same time, reaffirming Cartesian hierarchies of reason over passion is no longer neither viable nor desirable. The rationalist stance denies the affective dimension of human beings, including their empathic capacities. Neurobiologist Antonio Damasio's work (Damasio, 2021) confirms that there is no thinking without feeling. Thus, while artistic and photographic worlds may often simplify complex human and non-human realities, we believe that the expansive kind of vision proposed by art can still generate new moral imaginations of belonging and solidarity. As a privileged communicative form in contemporary culture, art has the power to disrupt dominant representations by crafting alternative narratives that transcend binary logic. Artists use imagination and storytelling to create spaces that are both intimate and political, enabling diverse perspectives to coexist while resisting the constraints of mobility regimes (Giudice & Giubilaro, 2015). As Edward Said (1979) has shown, geography is deeply enmeshed with struggles over ideas, forms, and modes of imagination.

As a starting point for the analysis of our selected artworks, we want to be mindful of the gap between seeing, caring, understanding, and acting in response to (the lack of) images that confront us with climate immobility injustices. Taking this awareness seriously, we ask: How can images of injustice spur viewers to action? Can artistic practices help us make sense of the present and the past? What forms of solidarity (should) we activate when, confronted with images that deepen our awareness of what human mistreatment and cruelty, we engage with the plight of victims with whom we do not share any political affinity? The ultimate answer to such questions may be this: it is our own responsibility as viewers to make the images speak, to actively give meaning to the images. Artists, curators, media, museums, and cultural organizations are in charge of the ethics of showing, but we are responsible for the ethics of seeing. The capacity to empathize with the oppressed and to grasp the structural underpinnings of climate and mobility injustice, as represented in contemporary art, resides not in the works themselves but within our psyches and bodies.

With this premise in mind, we propose the following claim as a guiding principle for the argumentation to follow: the most ethically captivating artistic practices addressing climate mobilities justice explore empathy as an aesthetic strategy, yet they allow moral ambivalence to be staged without attempting to resolve it. Rather than sidelining the empathic impulse or attempting to replace it with compassion, sympathy, empathic concern, or reason—as though such suppression or substitution were ever fully controllable, achievable, or beneficial in the first place—the case studies that follow explore strategies of perspective-sharing and historicization that offset the worrisome ambiguities of the empathic drive. These artistic practices function as sensitive research, provoking critical thinking and ethical reflection in its beholders. They not only invite us to empathize with fragile ecosystems and vulnerable communities; they also expose 'the hypocrisy demonstrated by the gap between discourses and practices that characterizes contemporary technoliberalism as the segment of the capitalist class that has poised itself as the one that can solve the crisis while actually increasing its profits and making it worse in the long term' (Terranova, 2024, p. 105).

From the Polis to the Cosmos

In the face of ecological and political challenges, can art contribute to the cultural debate so central in texts from Latour and Haraway to Baldwin and Sheller? Can art foster complicity with ecosystems and become an agent of change? How do we reclaim the cosmos as part of a broader wisdom? Can we follow the lead offered by artists who assert that their rights to belonging may begin in the polis but extend all the way out to the cosmos?

Nikos Papastergiadis (2012, 2023) a friend of many artists, museum curators, and cultural theorists) seeks to reframe our understanding of cosmopolitanism from the aesthetic perspective of the cosmos. Shedding light on how the philosophical tradition (from Kant to Habermas and Mouffe) 'focused on the rights to the polis, but lost sight of the cosmos' (Papastergiadis, 2023, p. 184), he argues that cosmopolitanism must be approached from the artist's point of view. Since the Enlightenment, cosmopolitanism has remained confined within a normative paradigm that dismissed Indigenous and peasant-based knowledge traditions as primitive or irrational. Aiming to reconnect cosmopolitanism 'to the spherical thinking that was evident in Antiquity and is emergent in ecological critiques, Indigenous cosmologies, and the artistic imaginary' (Papastergiadis, 2023, p. 153), he explores the bold interventions of artists who have pioneered new ecological and cosmological imaginaries. Claiming that modern art has extended not only the rights to the city (Lefebvre, 1968), but also the ancient rights to the cosmos, Papastergiadis explores the dynamics of hybridity and ambivalence, emphasizing how artistic practices carve out spaces of negotiation where new identities, perspectives, and futures can emerge. This hybridity mirrors the lived realities of climate-induced (im)mobilities, where displacement is not only a crisis but also a condition that shapes new cultural and political formations. In his view, 'love of cosmopolitanism is thus not a renunciation of the place you belong to. It is, on the contrary, a love for a place determined by the freedom to love all places' (Papastergiadis, 2023, p. 196). When brought to bear on mobility and migration, art's curiosity and creative energy, he contends, disrupts clear-cut binaries of belonging and alienation, stability and movement, the local and the global. This ambiguity, intrinsic to diasporic and displaced experiences, resonates deeply in artistic responses to climate change. As seas rise, forests burn, and landscapes become uninhabitable, artists challenge broad publics to rethink their position in an ecologically and geopolitically transformed world.

Many prominent contemporary artists, such as Emily Jacir,[2] Ai Weiwei,[3] and Tania Bruguera[4] can be seen as embodiments of aesthetic cosmopolitanism. In their work, we witness new expressions of long-standing cosmopolitanism principles such as hospitality, empathy, curiosity, and solidarity with others. By reworking everyday transactions and collaborating with local organizations, they resist the dominant political tendencies of their time such as unjust border controls, and mobility injustices, producing work that raises moral questions and prompts deep political engagement.

In *Where We Come From?* (2001–3) Emily Jacir, a Palestinian artist holding a US passport, asked thirty Palestinians living in the diaspora: 'If I could do anything for you, anywhere in Palestine, what would it be?' This simple yet profound question enabled Jacir to document why these people were unable to travel to locations in Israel, the West Bank, and the Gaza Strip, and to demonstrate that most of their requests were impossible to fulfill. In her artwork, we see her crossing physical and psychological borders and performing mundane yet emotionally charged acts on behalf of others—hugging a loved one, eating a favorite meal, watching the sunset—to reconnect Palestinian individuals with their homeland and each other.

In 2016, dissident Chinese artist Ai Weiwei covered the columns of Berlin's Konzerthaus, one of the German capital's best-known concert halls, with 14,000 life jackets collected on the Greek island of Lesbos. At the center of the stage stood a lifeboat inscribed with *#safe passage*. This act of solidarity, which sought to thrust the border crisis into public consciousness, continued in 2017 with the documentary *Human Flow*. The film's sweeping aerial shots and poetic imagery effectively convey

[2] https://darjacir.com/Team.

[3] https://www.aiweiwei.com/about.

[4] https://taniabruguera.com/.

the vast scale of the refugee crisis, emphasizing the interconnectedness of migration, climate change, and geopolitical instability. By documenting refugee experiences across 23 countries, Ai Weiwei critiques the systemic failures of wealthy nations at addressing the root causes of displacement, including war, environmental degradation, and economic inequality. Although some have critiqued his emphasis on shared humanity for lacking specificity and seemingly turning refugees into abstract figures, *Human Flow* remains a compelling call to action that challenges securitarian narratives portraying migration as a threat rather than a symptom of global injustice and inequality.

Similarly, Cuban artist Tania Bruguera—whose work draws inspiration from Indigenous resistance against Spanish colonialism—has created various performances resisting oppression, while opening an art school in Havana, and launching the *Immigrant Movement International* (2012) project at the UN in New York. During the *Atlas of Transitions|Home* festival held in Bologna in 2019, Bruguera developed a specific project entitled *Referendum*, which invited people to vote on the following proposition: 'Borders kill. Should we abolish borders?' For 10 days, voting stations were installed across the city, and a scoreboard displayed the daily results, reflecting the community's response and tracking participant choices. Here we may pause and ask: what is the impact of this *Referendum* from a legal point of view? None, of course. It was intended both as a performance and as a form of political activism. In this sense, art becomes a critical space for reflecting on the denial of rights during migration and on the geopolitical dynamics and exclusionary policies that shape mobility (Moralli et al., 2019). Art turns itself into a new public agora—an interactive space where diverse ideas, opinions, and insights on conflictual phenomena such as borders stimulate a renewed individual and collective responsibility in terms of co-citizenship (Balibar, 2002).

Through their politically engaged practices Jacir, Weiwei, and Bruguera may be considered practitioners of 'artivism,' a fusion of art and activism articulated at the intersection of aesthetic expression and resistance to societal injustices (Trione, 2022). Artivists situate art as a vital form of addressing pressing social challenges, positing alternative narratives that disrupt dominant discourses and give voice to marginalized experiences. Often operating outside traditional institutional boundaries, artivists provoke dialogue, foster participation, and critique hegemonic cultural practices. Ultimately, artivism embodies a multisensory subversion of mainstream imaginaries and centers participatory practices that critique inequalities and envision equitable futures. This approach underscores the transformative potential of contemporary art in reshaping collective imaginaries of climate-induced migration by fostering active participation in inclusive societies. Hence, art becomes not merely a reflection of reality but a medium for envisioning alternative futures grounded in care, resistance, and solidarity.

Engaging audiences is critical to countering distorted narratives about migration and climate change. Inspired by traditions such as 1960s experimental theater (Turner, 1982), many artists involve audiences through direct performance participation (Moralli et al., 2019), bodily engagement, or emotional reflection. This participatory approach redefines audiences as co-creators rather than passive observers, aligning with Augusto Boal's concept of the 'spect-actor' (Boal, 1979) and Jacques Rancière's idea of the 'emancipated spectator' (Rancière,2011). Within this dynamic space, spectators co-create meaning and oppositional narratives that upend dominant understandings of climate-induced migration.

Whether participatory or not, many of the selected artistic performances function as cultural-political interventions that disrupt hegemonic narratives of climate (im)mobilities. They open spaces for critical engagement, subverting mainstream discourses and crafting hybrid narratives that contest conventional norms of authority and inclusion (Martiniello, 2016). Among the artistic projects of the past decade addressing climate (im)mobilities, we focus on those that explicitly engage with climate (im)mobilities justice while inviting audiences to act or empathize from a radical purview. Many of these works prompt the beholders to know not only the human Other but also the non-human world. They expose toxic narratives about climate (im)mobilities while cultivating solidarities based on shared vulnerabilities and interdependencies. Although narrative frames often overlap and intermingle, these works foreground distinct political, ideological, and ethical implications tied to specific normative expectations.

Focusing on the four-fold framing of environmental migrants identified by Ransan-Cooper et al. (2015)—as victims, security threats, adaptive agents, and political subjects—the following sections examine how these conceptual frames are refined and enriched through the interventions of contemporary artists across the globe. Through the lens of their artistic practices, we observe how these frames are infused with new meanings and sometimes destabilized altogether, their boundaries rendered porous by the influence of decolonial artivism, contested legal personhood, and the tension between border abolitionism and adaptation-as-technofix. This reframing enables a richer understanding of climate (im)mobility that embraces the critical insights and provocations of artists and complements the awareness campaigns examined in the previous chapter. These artistic representations foster a renewed sensitivity—or perhaps a revitalized recognition—of alternative modes of being in the world that challenge underlying assumptions about interspecies relationships and temporalities and invite viewers to reconsider the ontological and ethical premises that inform current discourses on migration and the environment.

Victims, Who?

One key dimension of the artistic expressions analyzed in this chapter is their critical engagement with prevailing narratives of 'victimhood'—a deeply charged moral and political concept in contemporary discourse (Chouliaraki, 2024). As discussed in Chapter 4, media narratives around climate-induced (im)mobilities often present displaced populations through security-centric or humanitarian lenses, more or less inadvertently reinforcing power imbalances. While climate migrants from vulnerable regions are exposed to systemic injustices—including limited legal protections and exploitative labor conditions within rigid migration regimes—media coverage in Global North countries disproportionately emphasizes potential security threats over the structural inequities driving (im)mobilities. Lilie Chouliaraki's *Wronged: The Weaponization of Victimhood* (2024) critiques this dynamic by distinguishing between systemic suffering (endured by marginalized groups) and tactical victimhood (leveraged by privileged actors). Exploring the role of victim language in our society, Chouliaraki underscores how the victim is not a person per se, but rather a performative speech act through which the vulnerable self is produced at the very moment in which suffering is claimed. More specifically, focusing on the distinction between vulnerability and victimhood, Chouliaraki (2024, p. 6) denounces how the practice of claiming harm (of being considered a victim) has become the rhetorical province of the powerful: 'the proliferation of claims to victimhood produces its own victims by obfuscating truth—that is, by populating public discourse with too many voices of pain while selectively amplifying the voices of the already powerful over those of the underprivileged.'

Building on these insights, we can argue that dominant narratives often reduce (climate) migrants to abstract 'victims' of environmental forces, obscuring the colonial legacies and neoliberal policies that exacerbate their vulnerability. How, then, can art contest this framing of victimhood? How can artistic practices reveal the violence of non-recognition and subvert reductive binaries of vulnerability and agency?

Angela Tiatia,[5] a New Zealand-Australian artist of Sāmoan heritage, is the author of the performance video *Holding On* (2015), a poignant artistic response to climate displacement in Pacific Island communities, particularly in Tuvalu. *Holding On*[6] (2015) reconfigures the Pacific Islander experience of displacement, casting it not as passive suffering but as an act of resilience enmeshed in ancestral ecosystems. By collaborating with the ocean as a co-performer, Tiatia rejects Eurocentric 'disaster porn' narratives, instead positioning Indigenous communities as architects of adaptive futures. In *Holding On*, Tiatia lies on a cement slab as the ocean tide rises around her, symbolizing the struggle of Pacific Islanders to maintain their connection to their lands and cultures in the face of rising sea levels (Fig. 6.1).

The performance is both a personal and cultural indictment of climate change, reflecting Tiatia's own experiences and concerns about her ancestral homeland in Sāmoa, and illustrating the dynamic relationship between Indigenous peoples and their evolving environments. By using her body as a medium, Tiatia challenges traditional narratives of victimhood, displaying resilience and agency in the face of environmental challenges.

While colonial discourses flatten the complexity of mobility regimes with the singular term 'migration' and critique movement as indicative of underdevelopment and primitiveness, Tiatia's performance reclaims mobilities as a marker of perseverance, connectivity, and kinship. This work resonates with broader themes of climate migration by highlighting the intimate relationship between people and place, particularly in regions vulnerable to climate change. Tiatia's performance underscores the emotional, physical, and spiritual dimensions of holding onto cultural identity and land in the face of environmental degradation. It serves as a powerful metaphor for the challenges faced by communities that may be forced to migrate due to climate change, yet remain deeply connected to their ancestral lands. In the context of climate

[5] Meet the artist: https://www.youtube.com/watch?v=rit6rRrruG8.

[6] https://escholarship.org/content/qt8z45x79s/qt8z45x79s_noSplash_f5dc5481ceb4407dc094ec8ccc893106.pdf?t=rujat8.

Fig. 6.1 Angela Tiatia, Holding On, 2015, single-channel High-Definition video, 16:9, colour, sound. 12 minutes, 12 seconds, dimensions variable. Image courtesy of the artistartist and Sullivan+Strumpf. Courtesy of theartist artist Angela Tiatia

migration, *Holding On* encourages viewers to reflect on the human and cultural dimensions of environmental displacement, moving beyond simplistic narratives of victimhood or threat. It emphasizes the need for empathy and understanding in addressing the complex realities surrounding climate-induced migration, foregrounding the preservation of cultural heritage and community resilience. Her artistic practice offers a critical, decolonial intervention in the visual and discursive representations of climate change in Pacific Island communities, contesting the victimizing tropes often perpetuated by environmental documentaries and mainstream media. This approach resonates with contemporary decolonizing performance practices in Oceania, contributing to a broader movement that includes decolonial aesthetics and Indigenous futurism. By staging the ocean as a co-performer, Tiatia challenges prevailing portrayals of Pacific Islanders as inert sufferers of climate change. In this collaborative framework, both more-than-human elements and natural environments are given protagonism alongside local communities. This decentering move fosters a grounded imagination that repositions Indigenous peoples as active agents in co-creating their futures in concert with their environments. Tiatia's artistic strategy contributes to a growing body of scholarship that reconceptualizes Indigenous relationships with place, moving beyond reductive binaries of human versus nature. As such, Tiatia's work broadens the very definition of climate migration to encompass more-than-human actors and the environment itself, offering a holistic perspective on the impacts of climate change.

This work finds resonance in the transdisciplinary approach adopted by Andrea Conte (Andreco),[7] whose practice bridges environmental engineering and political ecology with visual art. Based in Rome, Andreco is a visual artist and researcher with a PhD and a postDoc in Nature Based Solutions for sustainable resource management and Climate Change adaptation and mitigation. Working at the crossroads of art, science, and political ecology, Andreco translates scientific data on desertification and sea-level rise into public art interventions that denounce the environmental racism produced by capitalist, anthropocentric, and patriarchal imperatives. In his large-scale installations—such as big flags and drapes carried by performers during artistic events (like festivals or the Biennale) or climate parades and activist rallies—Andreco represents only minerals, rocks, and water resources: non-human elements he identifies as exploited and oppressed by the Western anthropocentric system. 'The representation of deep geologies, rocks, boulders, land-slides, reefs, rivers, bacteria, viruses, plankton, single-celled organisms, everything that has hitherto been less considered by visual representation is a way of shifting attention to the non-human, generating affection, sympathy, complicity, and a sense of belonging' (Andreco, 2024, p. 100). Though his work is obviously meant as a provocation, it is also a creative way to restore dignity and give voice to normally silenced agents of the ecosystem. Through his work, Andreco reclaims art as a driver of ecological and epistemic justice, honoring the often-ignored chemical, physical, and biological processes that ensure life on Earth.

In dialogue with this wider ontological horizon, his art calls for systemic change, advocating for climate justice through multispecies mutualism and solidarity. As he writes, 'Bringing clarity and making interconnections between local ecosystem micro-dynamics and global geopolitical macro-issues is the key to a paradigm shift and turning towards another way of being in the world' (Andreco, 2024, p. 94).

[7] https://www.andreco.org/?lang=it.

In his *Displacement Malta*[8]—a site-specific version of the project realized for the opening of Malta Biennale of Art 2024—his performers carry flags and a long blue drape, enacting symbolic movements that evoke extreme meteorological events such as droughts, desertification, flooding, and sea-level rise. These flags represent part of Malta's coastline, cliffs, and sea landscapes. Celebrating subversive geography and transgressing borders, the performance challenges reductive frameworks of victimhood, invoking ecological, and multispecies solidarity in ways that echo Lilie Chouliaraki's critique of tactical victimhood. Through its choreographed 'anti-flags' and fluid scenography, the performance rejects security-centric or humanitarian tropes that render migrants abstract non-entities or passive victims of environmental forces (Fig. 6.2).

As a substitute, it highlights the systemic roots of climate injustice, such as the lack of legal protections for 'climate refugees'—an issue

Fig. 6.2 DISPLACEMENT MALTA is a site-specific version of Displacement (a collective performance about climate migration) realized for the opening of Malta Biennale of Art 2024. Courtesy of the artist Andrea Conte (Andreco), Climate Art Project. Photo: Daniele Cimaglia

[8] https://www.andreco.org/portfolio/displacement-malta-a-collective-performance-about-climate-migration/.

raised by Suzanne Maas, the coordinator of the *Friends of the Earth Malta's Climate Campaign*. The use of flags as fragmented maps and seascapes disrupts border logic, reframing migration as an ecological entanglement rather than a geopolitical threat. This subversive geography critiques the colonial and neoliberal policies identified as drivers of vulnerability, while the blue drape's ritualistic undulations evoke both environmental precarity and the fluidity of more-than-human networks. The performers—many of them climate activists—embody a form of 'agonistic solidarity' instead of victimhood, replacing an overemphasis on individualized pain with structural accountability (Chouliaraki, 2024).

Finally, the collective Mali Weil's[9] artistic practices are worth mentioning for their work reconfiguring cultural, social, and legal relationships between humans and other-than-human beings. Mali Weil is an artistic platform created by Elisa Di Liberato, Lorenzo Facchinelli, and Mara Ferrieri, based in Trento (IT). Through visual and performing arts, Mali Weil investigates the relationships between narratives and political imagination, creating multi-layered projects that range from performance, to design and film. Among its undertakings, the three-year project *Forests*[10] stems from Mali Weil's interest in the hidden history of the West, particularly in how neglected and uprooted pasts continue to shape its contemporary political imagination. *Forests* is conceived as a long-term transmedia journey spanning languages, media, and formats. The project comprises visual production, audio installations, design collections, writings, video works, and performances, while also adopting participatory structures such as temporary schools (either with kids or professionals), and curated educational programs. It unfolds as a polyphonic process of afforestation, bringing together artistic creation and scientific research to cultivate practices of 'social dreaming.' The research investigates forests as a conceptual, juridical, and political space that is deeply inscribed within Western imaginaries and narratives. As they explain in the press kit accompanying the film, 'It is in this very space that many of our contemporary juridical and political tools were born (for example the concept of private property as well as commons, legal personhood as well the contested

[9] https://www.maliweil.org/.

[10] https://www.maliweil.org/forests/.

concept of nature, but also more ephemeral skills such the ability of dreaming as a more-than-human diplomatic tool).'[11]

Mali Weil's *The Mountain of Advanced Dreams*[12] is a research-based art initiative that critically explores interspecies diplomacies and the possibilities of forging different relationships with other beings, living and non-living. Taking as its starting point a foundational relationship in the European imaginary—the one between humans and wolves—the project devises a fictional space that crosses multiple philosophical, legal, biological, and narrative directions. In reimagining a world where humans live in alliance with multispecies political communities, Mali Weil also launched the *School of Interspecies Diplomacies and Werewolfish Studies*, an open and itinerant time-space dedicated to training and research in interspecies diplomatic practices. This experimental school presents an innovative challenge to anthropocentric worldviews and invites audiences to imagine futures grounded in reciprocity rather than domination. Through storytelling and visual interventions that explore interspecies interactions, Mali Weil promotes co-evolutionary and reciprocal relationships—metaphors that resonate with the adaptive and collaborative nature of climate migrants. As with Tania Bruguera's *Referendum,*[13] the *School of Interspecies Diplomacy* holds no formal institutional or legal authority. It is 'just' an artistic gesture. At the same time, the School interrogates anthropocentric legal and cultural frameworks, advocating for 'ecocentric sensibilities' that dissolve entrenched binaries between humans and nature (Fig. 6.3).

These emergent representations recontextualize 'us' in settings in which climate migrants, along with non-human entities and environmental components, collectively assume central roles. This shift in narrative emphasis highlights the intricate dynamics between human and non-human participants within climate change discourse, revealing its profound implications for mobility and border politics. Thus, in a world shaped by climate instability, vulnerability is not the opposite of agency—just as agency itself is no longer the exclusive domain of human beings.

[11] https://www.maliweil.org/.

[12] https://www.maliweil.org/the-mountain-of-advanced-dreams/.

[13] https://bologna.emiliaromagnateatro.com/spettacolo/tania-bruguera-referendum-2/.

Fig. 6.3 Rituals|The Mountain of Advanced Dreams, Mali Weil, 2023, still da video. Courtesy Mali Weil e Museo Madre Napoli

In conclusion, Angela Tiatia, Andrea Conte (Andreco), and Mali Weil reconfigure climate (im)mobility justice through artistic practices that transcend national and anthropocentric frameworks, foregrounding a cosmic relationality that challenges dominant notions of security and sovereignty. Seen together, these artists re-sensitize the climate through visceral, embodied practices—Tiatia's endurance against rising tides, Andreco's choreographed floods, Mali Weil's ritualistic engagements—making ecological precarity palpable while recentering justice as a relational practice across species and generations. Their work connects Rancière's 'distribution of the sensible' (Rancière, 2009) with Indigenous cosmologies, re-routing climate discourse away from territorial securitization and toward pluriversal, justice-oriented struggles for coexistence within a shared planetary commons.

Reframing (in)Security

As explored in Chapters 3 and 4, for several decades narratives surrounding climate migration and their media representations have fluctuated between humanitarian and securitarian frameworks. The humanitarian discourse is often characterized by stereotypical portrayals of climate migrants, which foster and perpetuate notions of charity and compassion that are both dependent on and reinforce existing power imbalances (Musarò & Parmiggiani, 2022). In contrast, securitarian discourse encompasses narratives that exploit climate migration for ulterior purposes, as evidenced by the discursive and visual manipulations utilized by governments and institutions to justify military actions or enforce border closures (Musarò, 2017). This securitarian paradigm (Fassin & Pandolfi, 2010) is defined by the dehumanization and racialization of climate migrants, who are portrayed as looming threats to security, values, identity, culture, and resources. Such descriptions primarily serve to establish cognitive and social boundaries that legitimize the division between'us' and 'them' (Cappi, 2023).

Yet in the works of John Akomfrah[14] and Jason deCaires Taylor,[15] the securitarian frame is both subverted and reimagined. Rather than locating the threat within migration itself, their artistic interventions shift the locus of insecurity to the climate crisis per se. By reframing climate migration as the symptom of systemic ecological violence, their narratives seek to expose global injustices while redefining the concept of security within the context of a changing climate. Their artistic works shed light on colonial legacies and interspecies solidarity, thereby providing a more nuanced and critical perspective on the interplay between climate and mobility justice. This perspective proposes new definitions of security rooted in environmental justice rather than in a bordering arms race.

John Akomfrah is an artist and filmmaker whose works are known for their aesthetic investigations into memory, temporality, and the experiences of migrant diasporas from a global vantage. His artistic practice also consistently engages with themes of postcolonialism, ecology, justice, and the politics of aesthetics, as seen in his 2015 work *Vertigo Sea*[16] and his 2017 installation *Purple.*[17] In *Vertigo Sea* (2015), Akomfrah creates a haunting meditation on the oceans, delving into their intricate connections with war, migration, and the transatlantic slave trade. *Vertigo Sea* juxtaposes historical atrocities like the Zong Massacre with contemporary

[14] https://www.lissongallery.com/artists/john-akomfrah.

[15] https://underwatersculpture.com/.

[16] https://www.theguardian.com/artanddesign/2016/jan/25/john-akomfrah-vertigo-sea-bristol-arnolfini-london-lisson-gallery-auto-de-fe-the-airport-tropikos.

[17] https://magazine.tank.tv/tank/2017/10/john-akomfrah.

migrant crossings, positioning the ocean as both a graveyard of colonial exploitation and a site of ongoing climate displacement. This multidimensional narrative is further developed in *Purple*, a unique work that integrates nearly an hour of both new and archival footage from ten countries and various public archives, enhanced by narrations and an orchestral score. Through this aesthetic montage, Akomfrah underscores that environmental and social crises are not merely technological issues but the symptoms of malfunctioning of economic and social systems.

Akomfrah advocates for a radical approach to these challenges, suggesting that individuals assume the role of 'highwaymen' in the ongoing discourse. His chosen metaphor indicates the necessity for unconventional and potentially disruptive actions, with Akomfrah promoting the 'sequestration' of commodities that are ostensibly owned by others. This provocative idea reframes the redistribution of property as a potential solution to social and economic inequalities and injustices. Akomfrah's proposition highlights the urgent need for global collective action to prevent further environmental and social deterioration. By calling for immediate and widespread engagement with issues of climate and mobility injustice, his work offers a powerful critique of existing power structures. At the same time, it compels viewers to reflect on their own roles in perpetuating these dynamics. Through a temporal weaving of narratives, he reveals that the true existential threats lie not with migrants, but with the forces of racial capitalism and ecological collapse.

In a similar vein, Jason deCaires Taylor confronts securitizing narratives by shifting the focus to Indigenous solidarity, illustrating that the real security threat is environmental collapse, not migration. As a sculptor, environmentalist, and professional underwater photographer, Taylor has collaborated over the past 20 years with local communities around the globe, creating a series of Underwater Museums.[18] Each project transforms barren or damaged seabeds into flourishing habitats. These installations redefine our understanding of the underwater world by assigning it the same cultural and ecological importance as on-land museums, emphasizing the importance of preservation, conservation, and education. Taylor gained international recognition in 2006 with the creation of the world's first underwater sculpture park, situated off the West coast of Grenada in the West Indies. In 2016, he created

[18] https://underwatermuseum.com/.

Museo Atlántico,[19] a collection of over 300 submerged sculptures and architectural forms in Lanzarote, Spain, the first of its kind in European waters. Through a haunting visual dialogue between art and nature, the installations dramatize the tragic fate of asylum seekers traveling toward Fortress Europe, and highlight the social and political divisions within contemporary society (Giacomelli, 2023). Migrants are staged in a surrealist theatrical scenario composed of spectral human figures, the wall that halts their journey, and their dead bodies strewn about in a circle. At the same time, the works' immersion in a marine environment draws attention to the effects of climate change, creatively exposing the effects of ocean warming through world historical echoes of mobility.

In a similar way, *A World Adrift*[20] was launched in 2023 to emphasize the distinct vulnerabilities faced by Small Island Developing States (SIDS) due to climate change. Tragically, this installation—submerged 13 feet underwater off the coasts of Carriacou and Petite Martinique—has now become a poignant testament to these issues, as the impacts of climate change have manifested in real-time during the project's development. A *World Adrift* features an evocative collection of 30 boat sculptures, each helmed by a local school child, symbolizing the uncertain waters of the future. These young figures—cast as climate refugees—serve as a powerful reminder of the generational stakes involved in climate change but also the potential for defiance, solidarity, resilience, and hope (Fig. 6.4).

This project significantly reframes and reimagines narratives on climate (im)mobilities. By portraying children as navigators of uncertain futures, it disrupts dominant perceptions of climate migrants as passive victims, reframing them as agents of resilience and foresight. Through its artistic expression, *A World Adrift* invites audiences to engage with the realities of climate change and consider the broader implications for mobility justice and environmental stewardship. The installation not only raises awareness about the urgent issues confronting SIDS but also fosters a deeper understanding of how narratives can shape public discourse and influence policy responses to climate-induced displacement. Like all of deCaires Taylor's submerged installations, the work critiques extractivist

[19] https://underwatermuseumlanzarote.com/en/.

[20] https://underwatersculpture.com/projects/a-world-adrift/.

Fig. 6.4 A World Adrift. Courtesy of the artist Jason deCaires Taylor

systems responsible for climate collapse, in the process subverting anthropocentric paradigms by showing the interconnected fate of humans and ecosystems.

Both artists reject security-centric victimhood frameworks, instead aligning with the Black feminist ethos of *Undrowned: Black Feminist Lessons from Marine Mammals* (Gumbs, 2020), written by Alexis Pauline Gumbs. Employing a brilliant mix of poetic sensibility and naturalistic observation, the black American poet and activist explores the hidden link between the violence toward whales and enslaved ancestors, inviting us to reimagine a common survival through collective 'undrowning'—a practice of breathing and solidarity across species amid ecological violence. In their effort to show how the ecological crises are inseparable from the 'engineering' of racial hierarchies through slavery and plantation economies, Akomfrah's elegiac elephants in *Four Nocturnes*[21] and deCaires Taylor's coral-encrusted figures dissolving into reefs also echo the perspective developed by Malcolm Ferdinand in *Decolonial Ecology*

[21] https://wrightwood659.org/exhibitions/john-akomfrah-four-nocturnes/.

(Ferdinand, 2022). The ocean in Akomfrah's work becomes both a site of mourning and resistance, embodying Ferdinand's notion of the 'double fracture' that links colonial exploitation to environmental degradation by referring to the intertwined yet often kept separate histories of colonial domination and ecological destruction that define modernity. The *colonial fracture* arises from Western imperialism, and its imposition of racial hierarchies, slavery, and patriarchal systems, subordinating racialized and Indigenous peoples to white, male, Eurocentric power. On the other hand, the *environmental fracture* stems from the dualism of humanity and nature: insofar as technocratic and capitalist systems exploit ecosystems for profit, nature is necessarily treated as a resource rather than a living entity. Ferdinand argues that these fractures are not separate but deeply interconnected: colonial practices such as plantation economies and extractive industries simultaneously oppressed people and devastated ecosystems. In light of this entanglement, he advocates for a decolonial ecology—one that simultaneously addresses ecological and social injustices by envisioning alternative ways of inhabiting the Earth grounded in solidarity and mutual care among humans and non-human beings.

This approach challenges dominant environmentalist frameworks that often ignore the colonial roots of ecological crises, advocating for justice that is both historical and ecological. Similarly, Ellen Gallagher's *Watery Ecstatic*[22]—a series of mixed media drawings—explores a mythical 'Black Atlantis,' where descendants of enslaved Africans thrown overboard by their captors evolve into marine life, symbolizing resilience and transformation amid ecological and historical trauma. In alignment with Gumbs' reflections on marine mammals as models of collective survival and interspecies solidarity, Gallagher's work recenters the submerged Black Atlantic to reveal climate change itself as the product of centuries of exploitation, imagining a decolonial future in which migration no longer a threat, is reconceived as adaptive resilience.

Adaptation, but where?

Since the early 2000s, the discourse surrounding climate (im)mobilities has evolved significantly, transitioning from a focus on the coercive

[22] https://whitney.org/collection/works/18622.

dimensions of environmentally induced migration and associated security issues to a more nuanced understanding of migration as a viable and proactive adaptation strategy that requires governance and support (Piguet, 2013). In academic discussions, debates continue over the definition of concepts and terminology related to climate change adaptation and migration remains a prominent area of research (Vinke et al., 2020). Durand-Delacre (2023) usefully defines 'migration as adaptation' as the facilitation of labor migration—either internally or regionally—with the aim of enabling migrants to accumulate multiple forms of capital, including financial, human, social, and political resources, which can then be repatriated to their communities of origin. This process enhances adaptive capacity and may facilitate return migration. However, empirical evidence indicates a significant gap between the academic discourse on migration as adaptation and its practical application on policy, with Scott (2023) observing 'very limited evidence that states are adopting migration as adaptation policies.'

The artistic practices of Fijian illustrator Ropate Kama[23] contribute meaningfully to the migration-as-adaptation discourse by highlighting the proactive and positive contributions of climate migrants. Ropate Kama's work is part of the Climate Action Artist Residencies and explores the intergenerational psychological trauma caused by climate crises in indigenous communities. Through his residency, Kama blends Western and indigenous knowledge systems to create impactful narratives, such as his children's book *The Sun, the Sea, and Me*,[24] which uses poetic dialogues to address environmental issues and questions of relocation. As an Indigenous 'iTaukei' artist, Ropate integrates Western and Indigenous scientific perspectives in his book for children, imbuing it with motifs of healing while issuing a call to action against climate change. Illustrating how migrants can adapt and contribute positively to both their countries of origin and destination, the book narrates the climate crisis from the viewpoint of a young octopus called Rakūi, whose community decides to move due to climate-induced upheaval. Compelled to relocate, the protagonists are portrayed as creatively adaptive agents within their new environments. As mentioned above, the adaptation framework suggests that migration

[23] https://www.climateactionartistresidencies.org/residencies/residencies-ropate-kama.

[24] https://www.letsreadasia.org/read/d8fb0c93-0804-49ca-b1bd-D9222f161ba77?bookLang=4846240843956224.

can be a proactive strategy for coping with environmental challenges. However, this perspective also has its limitations, as noted by scholars who argue that migration often involves involuntary displacement rather than planned adaptation (Betts, 2013; Vinke, 2019).

In similar fashion, Carol Farbotko and Ingrid Boas developed a short animation for TED 2021 entitled *Where will you be able to live in 20 years?*[25] Focusing on the political consequences of generalized representations, the short video illustrates how rising extreme weather events are endangering coastal communities. Even for those with the resources to respond, the pace and nature of climate change makes adaptation increasingly prohibitive. Through the story of a family in Tuvalu, the animation conveys how climate change impacts human lives and mobility in far more complex, political, and heterogeneous ways than is often assumed. After undertaking efforts to adapt to their island and ensure their survival, the family is ultimately forced to relocate. Yet, they also demand the right to stay and the right to remain a nation. Rather than passively accepting to become 'climate refugees,' the video ends with the defiant slogan *We are not drowning, we are fighting*. As several authors (Farbotko et al. 2023; Kitara, 2023; Teafa, 2025) argue, the story of Tuvalu articulates a powerful claim to climate justice, urging an open and critical engagement with what mobility and immobility mean in the context of global warming.

Moving beyond representations of climate-affected people as mere migrants or passive victims, the video series produced by the HABITABLE[26] project examines how changes in the climate and environment shape how people perceive the 'habitability' of their places and influence their migration decisions. By developing plausible scenarios for how global warming may impact migration across different contexts, the series presents a range of characters employing diverse strategies for adaptation and survival—including, but not limited to, migration. These narratives underscore that the effectiveness of such strategies depends on a complex interplay of social, political, economic, environmental, and demographic factors. At the same time, they show that decisions to migrate are not

[25] https://ed.ted.com/lessons/where-will-you-be-able-to-live-in-20-years-carol-farbotko-and-ingrid-boas/think?dir=forward&question_number=3.

[26] https://habitableproject.org/scenario-narratives/.

taken lightly: while it can open up opportunities for migrants, their families, and their communities, it also entails significant costs. Moreover, the options, opportunities, and burdens for each person vary considerably depending on individual characteristics such as gender, age, and household position.

Despite these kinds of representational complexities, artistic responses to environmental crises play a vital role in raising awareness and fostering engagement with climate-induced migration. For instance, recent research on environmental art perception suggests that encountering climate-related artwork can increase support for climate policies by evoking emotional responses and cognitive processing (Leppänen & Aðalsteinsdóttir, 2025). Furthermore, studies on Nordic art and literature addressing environmental crises demonstrate how creative responses can activate local, regional, and global imaginaries, illustrating the complex interplay between art, activism, and environmental awareness. In this sense, the artistic work of Ropate Kama and the visual narratives discussed above contribute to broader debates in climate migration scholarship by illuminating the transformative potential of migrant communities. By framing migration as adaptation, these artistic interventions challenge prevailing narratives and emphasize the proactive contributions of migrants to both host and origin countries.

Broadening the Political 'we' of Climate (Im)Mobilities

A fourth dimension of the discourse on climate (im)mobilities broadens the conversation by incorporating mobilities that are often overlooked, such as internal and/or North-North movements, while also addressing the complexities of cross-border mobilities through the lenses of passport privilege and mobility justice. This perspective first challenges the dominant portrayal of climate migrants as solely impoverished individuals from the so-called Global South as it highlights the increasing displacement of Europeans resulting from the impacts of global warming. Secondly, it includes narratives that expose the challenges faced in cross-border movement due to climate-related factors, prompting a critical examination of how borders shape future climate realities. This framework also integrates emotional dimensions with collective narratives of justice, moving beyond the politics of pain and its accompanying affective responses—such as suffering, empathy, and anger. These narratives, despite their current marginalization, played a significant role in social change movements throughout the twentieth century (Chouliaraki, 2024).

Yinka Shonibare's art critically engages with themes of race, class, and the construction of cultural identity. Born in London in 1962 and raised in Lagos, Shonibare interrogates what he calls the 'tangled interrelationship between Africa and Europe'[27] with his British-Nigerian identity informing a sustained critique of colonial power structures. This decolonial ethos is emblematized in his installation *Refugee Astronaut* (2023),[28] exhibited as part of the Venice Biennale's *Foreigners Everywhere*[29] exhibition—the year's theme framing foreignness as both ubiquitous, mutable, and antithetic to fixed notions of identity. Shonibare's installation, positioned at the Arsenale entrance, confronts visitors with a paradox: the titular astronaut—typically a symbol of human ingenuity and progress—is rendered nomadic by a planetary crisis exacerbated by colonial capitalism. The work aligns with the exhibition's focus on Indigenous and diasporic perspectives, challenging Anthropocene narratives that universalize blame in order to erase historical accountability. *Refugee Astronaut* functions as a powerful visual metaphor for climate refugees, positioning them as political subjects and contesting conventional narratives of displacement and environmental crises. The sculpture features a life-sized astronaut adorned in African wax print fabric, carrying a netted sack of personal belongings. All these elements coalesce in the production of a visual that subverts colonial tropes of exploration and conquest.

Refugee Astronaut juxtaposes space exploration with forced migration, challenging Eurocentric visions of progress, and creating a broad 'we' where the whole human species risks becoming refugee astronauts. By reimagining the astronaut as a refugee, Shonibare reverses the conventional power dynamics associated with space exploration, portraying the figure as a nomadic survivor of ecological disaster rather than a conqueror of uncharted territories. This artistic intervention resonates with T.J. Demos' concept of radical Afrofuturism (Demos, 2016), which envisions alternative futures for African diasporic communities through speculative fiction and futuristic aesthetics. The *Refugee Astronaut* embodies

[27] www.yinkashonibare.com.

[28] https://wellcomecollection.org/articles/XYofFREAACQAp-Vl.

[29] https://www.labiennale.org/en/news/biennale-arte-2024-stranieri-ovunque-foreigners-everywhere.

a decolonial perspective by confronting the unjustness of the current mobility regime. By juxtaposing space-age imagery with the precarity of worldly displacement, Shonibare underscores how climate (im)mobility is not a 'Global South issue' but a shared planetary condition. As the Biennale's title asserts, 'foreignness' is a shared condition—one that demands dismantling the systems that make certain lives displaceable. Shonibare's sculpture thus becomes a call to 'undrown' together, forging futures where mobility is a right, not a problem.

Similarly, Lena Dobrowolska and Teo Ormond-Skeaping's[30] project, *You never know, one day you too may become a refugee*,[31] powerfully contributes to broadening the political 'we' in the discourse of climate (im)mobilities. The project revolves around a docudrama that imagines a speculative future in which climate change has triggered runaway displacement, including scenarios in which residents of the Global North are forced to seek refuge in the Global South. Through their work, the artists explore the entanglement of climate change with human rights and humanitarian concerns, using speculative narratives to highlight the shared vulnerability of all individuals to climate displacement, regardless of geographic or socioeconomic status. Inverting conventional migration routes, Dobrowolska and Ormond-Skeaping underscore the reality that anyone could become a climate migrant, thus confronting entrenched assumptions about who is at risk. By emphasizing universal precarity, their project fosters empathy and invites a more inclusive political conversation around climate migration. This artistic approach aligns with the idea of expanding the 'we' by encouraging viewers to recognize their own potential exposure to environmental dislocation. It humanizes migration by bridging the gap between distant, imagined others and the self, inviting audiences to engage with climate migration from a position of emotional and political proximity.

Moreover, by involving NGOs, policymakers, and broader publics through film screenings, workshops, and exhibitions, Dobrowolska and Ormond-Skeaping actively seek to transform cultural and political narratives around climate mobilityes justice. As the authors state in a panel during COP25 in Madrid—where their project received the 2019 COAL

[30] https://www.displacementjourneys.org/artists/lena-dobrowolska-teo-ormond-skeaping/.

[31] https://disasterdisplacement.org/news-events/coal-prize-winners/.

Prize recognizing artistic engagement with climate-related displacement—'We don't need more pictures of polar bears on thin ice. We need to connect the impacts of climate change to people.'[32]

Conclusions

Artistic engagement with climate change-induced (im)mobilities provides a vital framework for challenging dominant narratives and reimagining global interdependence. By creating innovative platforms for marginalized voices, especially from the Global South, the selected artistic expressions work to dismantle hegemonic Global North perspectives on climate migration (Mazzara, 2019; Musarò et al., 2022). Further, these artists challenge not only the separation between the metropolis and the colony but also the artificial divides between humans, animals, rocks, plants, and animals—inviting us to perceive the earth as an interconnected whole. More broadly, these artistic interventions contribute to a growing corpus that urges a shift in focus from border security to human and more-than-human safety when addressing climate-induced (im)mobilities, emphasizing the need for global solidarity and collective action in confronting the shared challenges posed by climate change.

As we have argued, these artists not only see 'their art as bridge-building exercises that connect different people' (Papastergiadis, 2023, p. 14); they also offer a more comprehensive and nuanced understanding of climate (im)mobility grounded in emotional and justice-oriented visions. Their art counters apocalyptic fatalism with the imaginative force of emancipatory utopias, showing that neither technological innovation nor policy alone will suffice. As Andreco (2024, p. 92) reminds us, 'We need a real cultural revolution that changes the Western perception of the planet and the perception of the other, of human and non-human otherness.'

Recalling the ongoing debate on empathy—its positive role as well as its limits in understanding others and responding to injustice—in this chapter we have seen how the selected artworks employ empathy as an aesthetic strategy. The efficacy of climate (im)mobility art lies not only in their diverse storytelling techniques but also in the ability of these forms to foster collective narrative resistance by politicizing the emotions

[32] https://disasterdisplacement.org/perspectives/in-conversation-with-lena-dobrowolska-and-teo-ormond-skeaping/.

they evoke aesthetically (Ahmed, 2013). Akomfrah's *Purple*, for instance, interweaves archival footage, landscapes, and different sound techniques to critique the entanglement of colonialism, ecological destruction, and displacement. By layering historical and contemporary images, he challenges linear narratives of crisis, illustrating how environmental degradation is not a new phenomenon but an extension of imperial violence. As Andreco (2024, p. 93) writes, 'Knowing the non-human, as well as knowing the other, the otherness that is often elected as the enemy, is the first step to creating empathy, a sympathy as Hume would say, that is necessary to preserve the ecosystem and ourselves.' Ultimately, however, the ability to empathize and grasp the roots of climate and mobility injustices lies not in the art itself, but in our own interpretative acts.

Through the conceptual frames of victimhood, (in)security, adaptation, and expanded political subjectivity, artists like Angela Tiatia and Andrea Conte (Andreco) contribute to a more nuanced understanding of climate (im)mobilities. As we have seen, Chouliaraki's critique of strategic victimhood underscores how dominant discourses obscure historical truths by amplifying elite claims to harm. Similarly, populist discourses weaponize victimhood to justify border securitization, casting host nations as 'victims' of migration influxes—a tactic that entrenches xenophobia. Tiatia's work highlights how systemic climate injustices are overshadowed by strategic appeals to victimhood that serve the interests of existing power structures. At the same time, we argued, Tiatia transcends false narratives of victimhood by providing a nuanced exploration of agency, adaptation, and cultural resilience in Pacific Island communities. Who qualifies as a (real) victim within the global climate change framework? How can (real) victimhood be reclaimed? Situated within the context of feminist environmental performance art in Oceania, her 'embodied' work provides a nuanced examination of loss and resilience in the face of rising sea levels and cultural erasure.

Likewise, Andreco's performances expand victimhood beyond anthropocentrism, foregrounding landscapes and waterways as agential forces, and position migrants in 'harmony with the ecosystem,' which reflects Chouliaraki's warning against narratives that sever human suffering from ecological violence. This approach echoes the storytelling currents of the Artist Circle for Climate Displacement which interconnect histories, affects, and material realities, as a form of resistance to the erasure of systemic respons-ability. Envisioning resilience through multispecies mutualism, Mali Weil's multimedia projects, and Andreco's *Displacement*

MALTA resonate with other projects that enact the new ecological politics by using transdisciplinary methodologies that foster new ways of imagining beyond an anthropocentric paradigm.

What if we welcome all beings—plants, animals, and things—to our parliament? This is the question that moves *The Parliament of Things*,[33] a speculative art research initiative inspired by Bruno Latour (Latour, 2018) that calls for the emancipation of animals, plants, and things. Among the project" many offshoots, the *Embassy of the North Sea*[34] acts as a diplomatic body representing North Sea-adjacent plants, animals, microbes, and people in legal and cultural forums. 'The Embassy of the North Sea is founded on the principle that the North Sea owns itself. As participants, we listen to, speak with, and negotiate on behalf of the sea and all the life that it encapsulates'.[35] Evoking more inclusive maritime perspectives, the *Embassy* investigates whether the North Sea should be an independent legal entity while raising questions such as: what is a human being in the Anthropocene? Where in climate change does the human end and the non-human begin? As Anne van Leeuwen, co-founder of the Embassy of the North Sea, states, 'In the Anthropocene we suddenly find ourselves on the same stage with so many non-human agents that question us with their interactions, forcing us to self-reflect'.[36] This is why the *Embassy of the North Sea* pleads for human rights that are no longer granted only to humans. Our rights should be co-terminous with the rights of air, water, and soil, the climate and biodiversity.

Art plays a crucial role in shaping different perspectives, offering visions of adaptation that resist both dystopian fatalism and techno-utopian solutions: Jason deCaires Taylor's underwater museums foster empathy by engaging local communities together with their history and culture, redefining human relationships with nature and conveying scientific information through art. His permanent site-specific sculptural works serve as both a memorial to lost homelands and an invitation to reimagine adaptation beyond mere survival and an embrace of creativity, renewal, and agency. Similarly, John Akomfrah's work rearticulates security concerns,

[33] https://theparliamentofthings.org/.

[34] https://www.embassyofthenorthsea.com/.

[35] https://www.embassyofthenorthsea.com/.

[36] https://theparliamentofthings.org/article/embassyofthenorthsea/.

highlighting climate change as the primary threat rather than migration, and advocating for global solidarity and collective action.

Showing how these artists elicit not only empathy but also a shared sense of respons-ability, our approach to the selected artworks has addressed politics as a configuration of a specific space where climate (im)mobility is not managed as a crisis but redefined through radical equality. Such a space is shaped by the encounter between the artist and the audience, who are jointly called upon to challenge the (often taken for granted) relationship between cultural imaginaries and political narrative.

Straddling the prerogatives of avant-garde transgression and decolonial care, contemporary artistic practices navigate—and productively destabilize—the ostensibly natural opposition between shock aesthetics and empathetic stewardship. The modernist avant-garde tradition, rooted in disruptive innovation and the 'shock of the new,' no doubt lives on in climate (im)mobility art. Vanguard acts aim to rupture complacency by forcing audiences to confront the visceral urgency of ecological collapse. Yet as the *Art for Radical Ecologies Manifesto* highlights, such strategies now coexist with Indigenous-informed practices of cosmic connectedness, and decolonial frameworks that recenter relational accountability, intergenerational wisdom, and non-human agency. While these approaches appear discursively contradictory (one prioritizing cathartic rupture, the other ancestral healing) their intersection manifests not as an incoherence but as a productive polyphony. We are aware, however, that both modes of expression risk institutional neutralization. The avant-garde's historical trajectory warns us of how transgression quickly turns into the mere acquisition of cultural capital. Similarly, decolonial aesthetics face appropriation through neoliberal 'inclusion' frameworks that romanticize care ethics while leaving extractive infrastructures undisturbed.

Ultimately, the Panicocene demands this dual stance: avant-garde urgency to dismantle old systems, paired with a decolonial cosmovisional approach that grafts new future imaginaries onto extant, failing knowledge. Both modes are essential if we want to avoid the risk of paralysis provoked by ecological emotions such as eco-anxiety, eco-worry, and eco-grief, which are associated with environmental degradation and the escalating climate crisis. In this liminal terrain, shock and care become not opposing forces but symbiotic tactics.

Finally, we want to highlight the transformative potential of artistic practices in reshaping and expanding methodologies across various sectors touched by climate (im)mobilities. While artists traditionally inspire action—a fundamental aspect of their practice—the unique methodologies, skills, and analytical frameworks cultivated through artistic endeavors offer invaluable contributions to this field. Many of the initiatives under consideration, certainly aim at amplifying marginalized voices; however, they also try to foster 'architectures of listening' that go beyond traditional politics by promoting solidarity in the face of growing inequalities (Macnamara, 2016).

Integrating artistic perspectives into broader multi-sector efforts addressing climate (im)mobilities requires intellectual flexibility and an ability to forge cross-disciplinary connections with researchers, media professionals, journalists, and activists. For example, Lena Dobrowolska and Teo Ormond-Skeaping's project underscores shared vulnerabilities, encouraging a broader recognition of climate migration as a universal risk, thus expanding the political 'we' to include all potential climate migrants. In a similar way, the researchers of the *Environmental & Climate Mobilities Network* cross academic boundaries by promoting a uniquely relational and transdisciplinary perspective. They highlight an emerging planetary politics of climate mobilities that contests the traditional division between states, land and water, the human and the non-human, and what we used to call mobile and immobile (Boas et al., 2024).

Beyond re-humanizing the experience of displacement, these artistic innovations also promote collective action and empathy, offering a hopeful vision of migration as adaptation and resilience in the face of environmental challenges. Such interdisciplinary collaboration can yield innovative solutions that transcend conventional boundaries while fostering a more comprehensive understanding of the complex interplay between climate change, human mobility, and societal responses. This integration can bridge concrete issues with grounded imaginaries across transnational contexts. Moreover, incorporating artistic methodologies into policymaking, scientific research, journalism, and community engagement could decolonize our approach to migration and climate change. This synthesis not only enriches public discourse but it can also cultivate more just and empathetic strategies for addressing a global issue requiring technical expertise and a profound human understanding.

References

Ahmed, S. (2013). *The cultural politics of emotion* (0th ed.). Routledge. https://doi.org/10.4324/9780203700372

Andreco. (2024). Art as tribute to the non-human. Exercises and open questions for Ecocentric expressions. In M. Baravalle, E. Braga, G. Riccio, & F. Timeto (Eds.), *Art for radical ecologies (manifesto)* (pp. 86–101).

Balibar, É. (2002). *Politics and the other scene*. Verso.

Baravalle, M., Braga, E., Riccio, G., & Timeto, F. (Eds.). (2024). Art for radical ecologies (manifesto). Bruno.

Barthes, R. (1982). *Camera lucida: Reflections on photography* (26. print). Hill and Wang [u.a.].

Betts, A. (2013). *Survival migration: Failed governance and the crisis of displacement*. Cornell University Press.

Bloom, P. (2018). *Against empathy: The case for rational compassion (first published in vintage)*. Vintage.

Boal, A. (1979). *Theatre of the oppressed*. Pluto Press.

Boas, I., Farbotko, C., & Bukari, K. N. (2024). The bordering and rebordering of climate mobilities: Towards a plurality of relations. *Mobilities, 19*(3), 521–536. https://doi.org/10.1080/17450101.2023.2279095

Braga, E. (2024). Art as tribute to the non-human. Exercises and open questions for Ecocentric expressions. In M. Baravalle, E. Braga, G. Riccio, & F. Timeto (Eds.), *Art for radical ecologies (manifesto)* (pp. 30–43).

Breithaupt, F., & Hamilton, A. B. B. (2019). *The dark sides of empathy (1st edition)*. Cornell University Press.

Cappi, V. (2023). *Immaginare l'altrove nell'epoca dell'Antropocene. Media, confini e cambiamenti climatici*. Franco Angeli.

Chouliaraki, L. (2006). Towards an analytics of mediation. *Critical Discourse Studies, 3*(2), 153–178. https://doi.org/10.1080/17405900600908095

Chouliaraki, L. (2013). *The ironic spectator: Solidarity in the age of post-humanitarianism*. Polity Press.

Chouliaraki, L. (2024). *Wronged: The weaponization of victimhood*. Columbia University Press.

Cucuzzella, C. (2021). Making the invisible visible: Eco-art and design against the Anthropocene. *Sustainability, 13*(7), 3747. https://doi.org/10.3390/su13073747

Damasio, A. R. (2021). *Feeling and knowing: Making minds conscious*. Knopf Doubleday Publishing Group.

Demos, T. J. (2016). *Decolonizing nature: Contemporary art and the politics of ecology*. Sternberg Press.

Durand-Delacre, D. (2023). Climate-related mobility mostly occurs within Borders: Disentangling factual truth from a convenient narrative. *Researching Internal Displacement.*. https://researchinginternaldisplacement.org/short_pieces/climate-related-mobility-mostly-occurs-within-borders-disentangling-factual-truth-from-a-convenient-narrative/

Farbotko, C., Boas, I., Dahm, R., Kitara, T., Lusama, T., & Tanielu, T. (2023). Reclaiming open climate adaptation futures. *Nature Climate Change, 13*(8), 750–751. https://doi.org/10.1038/s41558-023-01733-1

Fassin, D., & Pandolfi, M. (Eds.). (2010). *Contemporary states of emergency: The politics of military and humanitarian interventions*. Zone Books.

Ferdinand, M. (2022). *Decolonial ecology: Thinking from the Caribbean world* (A. P. Smith, Trans.). polity.

Giacomelli, E. (2023). *Panicocene: Narrazioni su cambiamenti climatici, regimi di mobilità e migrazioni ambientali*. Franco Angeli.

Giudice, C., & Giubilaro, C. (2015). Re-imagining the border: Border art as a space of critical imagination and creative resistance. *Geopolitics, 20*(1), 79–94. https://doi.org/10.1080/14650045.2014.896791

Gumbs, A. P. (2020). *Undrowned: Black feminist lessons from marine mammals*. AK Press.

Kaplan, E. A. (2011). 15 empathy and trauma culture: Imaging catastrophe. In A. Coplan & P. Goldie (Eds.), *Empathy: Philosophical and psychological perspectives* (pp. 255–276). Oxford University Press. https://doi.org/10.1093/acprof:oso/9780199539956.001.0001

Khosravi, S. (2011). *"Illegal" traveller: An auto-ethnography of borders (1. publ. in paperback).* Palgrave Macmillan.

Kitara, T. (2023). Tuvaluans need action, not pity. *E-Tangata.*. https://e-tangata.co.nz/commentand-analysis/tuvaluans-need-action-not-pity/

Latour, B. (2018). *Down to earth: Politics in the new climatic regime*. (English edition). Polity.

Lefebvre, H. (1968). *Le droit à la ville*. Economica-Anthropos.

Leppänen, K., & Auður Aðalsteinsdóttir (Eds.). (2025). Creative responses to environmental crises in Nordic art and literature. .

Macnamara, J. (2016). Organizational listening: Addressing a major gap in public relations theory and practice. *Journal of Public Relations Research, 28*(3–4), 146–169. https://doi.org/10.1080/1062726X.2016.1228064

Martiniello, M. (Ed.). (2016). *Multiculturalism and the arts in European cities* (1. issued in paperback). Routledge.

Mazzara, F. (2019). *Reframing Migration: Lampedusa, Border Spectacle and the Aesthetics of Subversion* (P. Antonello & R. S. C. Gordon, Eds.; 1st, New ed ed.). Hachette livre. doi:https://doi.org/10.3726/b15214.

Meskimmon, M. (2023). Transnational feminisms and art's transhemispheric histories: Ecologies and genealogies. *Routledge, Taylor & Francis Group*. https://doi.org/10.4324/9780429507816

Mirzoeff, N. (2014). Visualizing the Anthropocene. *Public Culture, 26*(2), 213–232. https://doi.org/10.1215/08992363-2392039

Moeller, S. D. (1999). *Compassion Fatigue: How the Media Sell Disease, Famine, War and Death* (1st ed.). Routledge. https://doi.org/10.4324/9780203900352

Moralli, M., Musarò, P., Paltrinieri, R., & Parmiggiani, P. (2021). Creative resistance. Cultural practices, artistic activism and counter-hegemonic narratives on diversity. *Studi culturali, 2*, 163–180. https://doi.org/10.1405/101880

Moralli, M., Musarò, P., & Parmiggiani, P. (2019). Borders kill. Tania Bruguera's referendum as an artistic strategy of political participation. *Journal of Mediterranean Knowledge, 4*, 137–159. https://doi.org/10.26409/2019JMK4.2.10

Musarò, P. (2017). Mare nostrum: The visual politics of a military-humanitarian operation in the Mediterranean Sea. *Media, Culture & Society, 39*(1), 11–28. https://doi.org/10.1177/0163443716672296

Musarò, P., Papastergiadis, N., & Peja, L. (2022). Reimagining narratives on migration. The role of media, arts and culture in promoting transcultural dialogue. *Comunicazioni Sociali, 1*, 3–16. https://doi.org/10.26350/001200_000154

Musarò, P., & Parmiggiani, P. (2022). *Ospitalità mediatica*. Le migrazioni nel discorso pubblico.

Paltrinieri, R., Parmiggiani, P., Musarò, P., Moralli, M., & (Eds.). (2020). *Right to the city, performing arts and migration*. FrancoAngeli.

Papastergiadis, N. (2012). *Cosmopolitanism and culture*. Polity Press.

Papastergiadis, N. (2023). *The cosmos in cosmopolitanism*. Polity Press.

Piguet, E. (2013). From "primitive migration" to "climate refugees": The curious fate of the natural environment in migration studies. *Annals of the Association of American Geographers, 103*(1), 148–162. https://doi.org/10.1080/00045608.2012.696233

Rancière, J. (2009). *The politics of aesthetics: The distribution of the sensible* (Reprinted). Continuum.

Rancière, J. (2011). *The emancipated spectator (paperback ed)*. Verso.

Ransan-Cooper, H., Farbotko, C., McNamara, K. E., Thornton, F., & Chevalier, E. (2015). Being(s) framed: The means and ends of framing environmental migrants. *Global Environmental Change, 35*, 106–115. https://doi.org/10.1016/j.gloenvcha.2015.07.013

Ross, C. (2022). *Art for coexistence: Unlearning the way we see migration*. The MIT Press.

Rumford, C. (Ed.). (2014). *Citizens and borderwork in contemporary Europe* (1. issued in paperback). Routledge.

Said, E. W. (1979). *Orientalism*. Vintages books.

Scott, M. (2023). Adapting to climate-related human mobility into Europe: Between the protection agenda and the deterrence paradigm, or beyond? *European Journal of Migration and Law, 25*(1), 54–82. https://doi.org/10.1163/15718166-12340144

Seiger, F., LeLouvier, K., Jinkang, A., Granger, S., Villani, L., Sbaa, M., Cappi, V., Bermejo, R., Puzzo, G., Shubin, S., & Bratskas, R. (2022). The PERCEPTIONS handbook. *A guide to understand migration narratives & addressing migration challenges.*. https://www.perceptions.eu/handbook/

Singer, P. (2011). *The expanding circle: Ethics, evolution, and moral progress (1st Princeton University press pbk ed)*. Princeton University Press.

Smith, T. (2011). *Contemporary art: World currents*. Prentice Hall.

Sontag, S. (2013). *Regarding the pain of others*. Penguin Books Ltd..

Teafa, L. (2025). Unpacking climate migration. *Pakhuis de Zwijger.*. https://www.facebook.com/pakhuis.dezwijger/videos/unpacking-climate-migration/572865665663833/

Terranova, T. (2024). Technoliberalism and the network social. *Theory, Culture & Society, 41*(7–8), 89–104. https://doi.org/10.1177/02632764241299388

Trione, V. (2022). Artivismo: Arte, politica, impegno.. Giulio Einaudi editore.

Turner, V. W. (1982). *From ritual to theatre: The human seriousness of play*. Performing Arts Journal Publications.

Vinke, K. (2019). *Unsettling settlements—cities, migrants, climate change: Rural-urban climate migration as effective adaption?* LIT.

Vinke, K., Bergmann, J., Blocher, J., Upadhyay, H., & Hoffmann, R. (2020). Migration as adaptation? *Migration Studies, 8*(4), 626–634. https://doi.org/10.1093/migration/mnaa029

Yumagulova, L., Parsons, M., Yellow Old Woman-Munro, D., Dicken, E., Lambert, S., Vergustina, N., Scott, J. C., Michell, P., & Black, W. (2023). Indigenous perspectives on climate mobility justice and displacement-mobility-immobility continuum. *Climate and Development, 1–18*. https://doi.org/10.1080/17565529.2023.2227158

CHAPTER 7

Conclusions. It's the End of the World as we Know it

Cosmovisional Justice and Decolonial Narratives

> 'Maybe we're the new dinosaurs and need to watch out. But then maybe against all the odds we'll migrate to Mars where we'll start a colony of gentle preservers, people who'll want to keep the red planet red, we'll devise a planetary flag because that's a thing we lacked on earth and we've come to wonder if that's why it all fell apart, and we'll look back at the faint dot of blue that is our old convalescing earth and we'll say, Do you remember? Have you heard the tales? Maybe there's another parent-planet–Earth was our mother and Mars, or somewhere, will be our father. We are not such orphans-in-waiting after all'
> —Samantha Harvey (2023 p. 165)

Samantha Harvey's 2023 novel *Orbital* follows six fictional astronauts for over 24 hours aboard an orbiting space station. Winner of the 2024 Booker Prize, it interweaves the astronauts' official duties with deep reflections on humanity and climate change. As eco-fiction, Harvey's novel does not dwell on graphic depictions of future catastrophes; rather, it offers a panoramic, almost reverent view of Earth's fragile ecologies. From that lofty vantage, her characters are struck by the irony that we—the planet's self-appointed caretakers—are the ones systematically destroying it. Why are we doing this?

E. Giacomelli and P. Musarò, *Climate Mobility Justice*,
https://doi.org/10.1007/978-3-032-07907-7_7

Through its twists and turns, this book has explored the complex interplay between climate change and human (im)mobility, critically examining the dominant narratives and representations that shape our understanding and responses to these interlinked phenomena. Through diverse perspectives and analyses, we have demonstrated how extant discourses often perpetuate harmful stereotypes, foster panic-driven reactions, and reinforce existing power imbalances. By examining the mechanisms of climate (im)mobilities discourse through multiple lenses, we have exposed how dominant narratives often perpetuate colonial power structures while obscuring indigenous cosmovisions that offer alternative pathways. We interweaved storytelling methodologies with climate justice frameworks and demonstrated that *who* tells climate mobility stories—and *how* they tell them—fundamentally influences our collective response to one of humanity's greatest challenges. Addressing the climate mobility nexus, we found, requires not just policy reform but a radical shift in how we conceptualize and communicate its narrative and visual politics. In other words, we need better stories to make better policies.

Throughout these pages, we exposed how mainstream narratives of climate-induced movement swing between two extremes—apocalyptic mass displacement and techno-utopian solutions—both of which ignore underlying social inequalities. Such framings frequently cast affected communities as passive victims—stripped of agency and historical context—while eliding the differentiated responsibilities for the impacts of climate change. The dominant metaphors and language employed—from 'climate refugees'to 'waves' and 'floods' of migrants—continue to fuel the process of othering and legitimize the militarization of the borders by framing people on the move as threats rather than individuals exercising their right to mobility and safety. Tellingly, when climate change-related movements of people occur in the Global South, the media, international organizations, and even researchers will often refer to them as 'climate migrants.' In contrast, similar events within Europe—take, for instance, the floods in Emilia-Romagna (Italy), which displaced over 40,000 residents, or the extreme heatwaves and hailstorms that struck Southern and Eastern Europe in the summer of 2023—are typically framed in a very different light. In these cases, the media and policymakers refer to those affected as climate-displaced 'citizens' or displaced European 'tourists' (Boas et al., 2024) (Fig. 7.1).

The politics of climate mobilities demands we scrutinize how climate-related human movement is represented and framed. This includes

Fig. 7.1 The center of Traversara (Ravenna) flooded due to the breach of the Lamone River embankment, Sept. 19, 2024. Ph. Michele Lapini. Courtesy of the photographer Michele Lapini

being mindful of how policies, activism, news media, and documentaries construct narratives that make climate mobilities visible and intelligible. What are the consequences of different forms of representation? What do they reveal, obscure, or exclude? Climate mobility narratives are selective representations that prioritize certain elements—problems, victims, solutions—while silencing others, often reinforcing existing hierarchies of domination and control. These frames contribute to the dehumanization of those most affected, reducing complex realities to simplistic binaries while elevating a 'politics of pity' over one of systemic justice (Boltanski, 1999; Chouliaraki, 2013).

By historicizing 'crisis exceptionalism,' we have dislodged its perception as a series of discrete catastrophes and repositioned climate (im)mobility along a continuum of human movement (Farbotko, 2022). This perspective shift acknowledges the complex interplay of environmental, political, economic, and social factors shaping human movement in the context of climate change. We have underscored migrant agency and backdropped victimhood tropes. We have unveiled how our pervasive sense of panic and uncertainty about global warming and the mass movement of people is not simply a natural response. Rather, it is a manufactured reaction stemming from a toxic narrative disseminated over the past few decades by actors who have no interest in actually changing our contemporary social, political, and economic system. This narrative often fuels reactive, short-term solutions, while hindering more thoughtful, long-term approaches aimed at tackling the root causes of the crisis. By unveiling the inner workings of this narrative, we called for a shift away from panic-driven responses toward more grounded, hopeful, and transformative actions.

Narrative is not merely representational; it can also resist and regenerate. We sorely need frameworks capable of reimagining more equitable and sustainable relationships between human and more-than-human communities. This volume is, above all, a deliberate act of imagination—a practice of world-building premised on the conviction that envisioning alternative futures is essential to realizing them. Moral imagination is integral to our critical reasoning capacities, and the ability to forecast future scenarios plays an important role in orienting our moral reflection (Draper, 2024). By reflecting on the future implications of climate mobilities, we sharpen our ability to make judicious decisions about present-day responsibilities and actions. Aware of how exceptionally complex and ongoing the construction of new worlds is, we offer this work, with all its limitations and inherent incompleteness, as a provisional toolbox. Our hope is that it may animate a broader dialogue, provoke critical reflection, and inspire collective imaginaries of more just and livable futures. To this end, we have included analyses of awareness campaigns and artistic practices as communicative acts that can be used to either confirm or challenge mainstream narratives. Many of these interventions envision meaningful futures beyond dystopia or denial. Though differing in terms of geographical context, temporality, and the story roles of their protagonists, they are narratives that 'cope with crisis.' By integrating multi-scalar justice with art, we cast climate (im)mobility justice as a multidimensional framework: one that addresses systemic inequities while envisioning adaptive resilience beyond borders. Artistic interventions not only critique existing paradigms but also empower audiences to reimagine global relationships and responsibilities in settings of climate change-induced mobility.

In her essay *On Heat* (Ferreira da Silva, 2018), theorist and artist Denise Ferreira da Silva asks: 'How to unthink the world? How to release it from the procedures and tools that presume everything that exists or happens is an expression of the human? What descriptors will support critical accounts of climate change that do not—like the theses on the Anthropocene—simply reaffirm the human and its unique attributes? What accounts of change can displace Universal time and release its grip on our imagination?'.[1] The challenge here is to imagine change beyond the normative story of modern Europe's march of progress—unfolding first through Natural Law, and then via natural selection and economics—as a stand-in for the universal truth of humanity. Asking us to move toward an account of material transformation, Ferreira da Silva calls for 'a radical shift in thinking that artistic practice is more equipped to mediate because its craft cannot but attend to the imagination.' Reminding us that rising temperatures may provide a guide for thinking through how colonial, racial, and capitalist processes are deeply—and materially—co-implicated, Ferreira da Silva highlights that artists play an important role in weaving alternative visions, vocabularies, and narratives capable of upending 'too late' and end-of-times discourse while experimenting with alternative economies, values, and relational forms.

We have thus included an analysis of artistic practices whose central focus is the connection between creativity and the cosmos. The experience of art can lead us to rethink the cosmopolitan imaginary, enriching the old Eurocentric vision with decolonial narratives of contact and exchange. As Papastergiadis (2023, p. 154) writes: 'If our vision of cosmopolitanism is kept within philosophical parameters of normative cosmopolitanism, then the sphere of the cosmos will contract.' From Saint Paul's claim of universal brotherhood to the Enlightenment's vision of international cooperation, the idea of cosmopolitanism has steadily moved closer to the realm of ethics and politics. Compared to its foundation in ancient Greece with the Stoics, however, the scope of cosmopolitan thinking has been significantly reduced. The modern philosophical tradition, with its focus on the rights of the *polis*, lost sight of the cosmos—and since the Enlightenment, it has also excluded Indigenous and peasant-based traditions of knowledge long deemed primitive or irrational. This rejection of the spherical connection between the political/ethical and the cosmological/aesthetic has led to a contraction into a narrowly normative paradigm

[1] https://canadianart.ca/features/on-heat/

of worldliness—one that sets us up for further frustration and disillusionment. Cosmopolitanism, in its fullest sense, entails both (1) a way of being in the world that aspires to universal moral connectedness and (2) an emergent social order that extends political rights beyond exclusivist territorial boundaries. Yet to feel that individual sense of moral connectedness and to organize collective modes of solidarity, we must employ a mix of both aesthetic and deliberative modes of comprehending and evaluating cultural similarities and differences. As we have shown in earlier chapters, normative cosmopolitanism is driven by moral imperatives, whereas aesthetic cosmopolitanism is propelled by the creative constitutive. Rather than passively accepting the policing mechanisms of nation-states and their militarized borders, aesthetic cosmopolitanism offers a different optic on human nature—one that embraces our impulses for care, curiosity, and creativity. Thus, according to Papastergiadis (2023, p. 186), 'the creative constitutive *in* cosmopolitanism precedes the moral imperative *for* cosmopolitanism.'

If we judge the viability of cosmopolitanism solely on the terms and scale of the nation-state, we will always miss its essence—it simply does not reside there. To truly grasp the cosmopolitan ethos, we must look at it from another angle, which is necessarily above and below the nation. A more nuanced scalar framework is one that includes everyone and everything, not only 'my people,' and a pluriversal planetary perspective reclaims the cosmos as part of a more far-reaching wisdom.

For this reason, we supplemented our investigation on climate (im)mobilities justice with a decolonial approach that demands the dismantling of epistemic hierarchies and the centering of voices, experiences, and knowledge marginalized by colonial modernity. A decolonial perspective confronts the dominance of Eurocentric, civilizing worldviews that have long claimed to be the sole arbiters of progress and rationality. In contrast, the cosmocentric philosophies of many Indigenous traditions—now embraced by a growing body of scholars and activists (Krenak, 2024; Mignolo, 2000; Yumagulova et al., 2023; Quijano, 2007)—offer alternative ethical, legal, and political frameworks grounded in relationality, reciprocity, and respect for all life forms. This perspective illuminates how extractive capitalism—rooted in colonial patterns of land appropriation and resource exploitation—remains a driving force behind the climate crisis and its disproportionate impacts on historically colonized communities in the Global South. Rejecting anthropocentric rationalism, these forms of knowledge production— indexed as 'epistemologies of the South' by Boaventura de Santos (2010)—critique the commodification of nature under free-market logic. Under this framework, climate justice is not merely a legal or political issue but a moral imperative that acknowledges the systemic violence inflicted upon the Earth and its more-than-human inhabitants. Through this cosmovision, the climate crisis is understood as a relational and spiritual injustice, requiring an ethical response rooted in interdependence, earthly reverence, and collective respons-ability.

This means moving beyond the political and economic logics that prioritize legal-financial interests over socio-environmental priorities. It means resisting the reduction of climate mobility to mere data points or crisis imagery and, instead, cultivating narratives that are affective, embodied, and collective. As scholars, journalists, and activists, we must deconstruct potentially harmful discourses while nurturing emancipatory narrative frameworks capable of sustaining what Lear (2006) terms 'radical hope' in the face of unprecedented environmental and social transformations. We need to rethink our moral and political respons-abilities—and the shape that justice takes at different scales.

Consider the case of Tuvalu and other island nations, where sea-level rise may result not only in the loss of land—which could lead to a substantial rebordering of its territory—but also in the dispersion of cultural identity and attachment to place. Due to a recent glacier melt in the Alps, Italy and Switzerland had to formally move their shared border (Boyd, 2024). In the Arctic, shrinking sea ice has sparked a scramble for geopolitical advantage, as the ongoing threats to invade Greenland vividly demonstrate. As Boas et al. (2024) highlight, these examples show how environmental changes cut across divides that we often take for granted—between states, land, and water, human and non-human, mobile and immobile.

In sum, an emerging planetary politics of climate mobilities calls for a multi-scalar approach to justice—one that recognizes the entanglement of bodies, infrastructures, and planetary systems; that confronts the colonial and racialized architectures of movement; and that supports grassroots resistance, sanctuary jurisdictions, and the radical hope of climate justice movements. The challenge before us is not only to dismantle the walls and detention camps of the present, but to reimagine how all boundaries—human and non-human, regulatory and territorial—shape our societies, world politics, and our relationship with the environment (Boas, 2025; Sheller, 2018).

Building on our critical reflection of cosmopolitanism and creativity, we propose a pluriversal cosmovisional justice that centers diverse ways of knowing, being, and relating to the world-especially those rooted in Indigenous and non-Western traditions. This perspective challenges the anthropocentric and extractivist paradigms that underpin much climate discourse. By foregrounding relationality—between humans, non-humans, and the land—cosmovisional justice insists on the sacredness and interconnectedness of all life. This perspective disrupts the colonial logic that frames nature solely as a resource and migration as a threat. Instead, it invites us to see climate (im)mobilities not as isolated crises but as symptoms of deeper historical injustices and ruptures in our relationship with the Earth, recognizing the plural dimensions of justice at the level of the individual, the locality, the region, the nation, and the planet.

Moving beyond the Panicocene

In recent decades, an abstract notion of the 'enemy' has been wielded politically to justify ever-stricter policies and the erosion of individual liberties. Due to this shift, security is no longer a careful balancing act of protection and liberty but has become a binary struggle between a normative 'us'—the law-abiding citizenry—and a menacing 'them,' whether migrants, racialized others, or perceived criminals.

Fear now permeates political discourse both as an outward response to constructed threats and as an internal mechanism for managing domestic inequalities. As Corey Robin (2004) observes, political fear has this dual orientation: it targets external enemies while deflecting attention from structural injustices and socioeconomic conflict. And, as Nussbaum (2018, p. 26) reminds us, 'fear makes us want to avoid disaster. But it certainly does not tell us how.' Power's true efficiency lies in converting that vague desire to escape terrible harm into a rationalization for internal control. This dynamic fosters what may be termed a 'culture of enmity,' wherein marginalized populations—such as refugees, the unemployed, or ethnic minorities—are positioned as risks to social cohesion. In periods of heightened social stress or institutional fragility, these groups are turned into scapegoats. René Girard's (1972) classic theory of scapegoating continues to elucidate this process: during times of collective anxiety, societies seek a substitute victim whose symbolic neutralization restores a fragile sense of order. Such scapegoating mechanisms are often symptomatic of deeper systemic failures. Robert Castel (2011) calls this 'security frustration'—a disjuncture between institutional narratives that promise comprehensive protection and the actual, often limited, capacity of those institutions to deliver on their promises of security. Rather than addressing the structural roots of social precarity, political responses tend to displace public anxiety onto the most vulnerable, reinforcing exclusionary practices under the guise of restoring order.

Societies typically respond to growing insecurity and social atomization in one of two ways. The first is repression: extending criminal laws into every corner of public life, empowering police to act coercively, and drawing rigid divides between 'legitimate' and 'illegitimate' populations. This approach is cheap—basic needs can be denied to those labeled as enemies of the social order—and politically expedient since scapegoating generates media spectacle and electoral popularity.

The second—but far more demanding—strategy tackles the root causes of precarity. It combines prevention, care, and reintegration with reforms in education, social services, and healthcare to address the suffering and destructive behaviors that arise from exclusion. This path is slow, costly, and devoid of immediate electoral spoils. Because this non-magical formula always presupposes its practitioners as implicated in social problems as individuals and as institutions, it does very little to provide immediate catharsis. It is a frankly dull show which cannot be easily spectacularized, but it does build genuine resilience, rather than perpetuating cycles of punishment and stigma.

We inhabit systems that manufacture material insecurity. Just two decades ago, Robert Castel (2011) noted that we live in the safest societies in human history—and yet feel more insecure than ever. How can this paradox be explained? Insecurity has become fertile ground for power in an age of rampant individualism and fear. Crucially, security itself carries a dual meaning: it can justify control, surveillance, and repression, or it can inspire protection, vigilance, care, and solidarity.

The concept of a 'moral panic' as originally formulated by Stanley Cohen (2011), describes a social phenomenon wherein a person, group, or condition is perceived as a threat to societal values and interests, leading to an exaggerated and often irrational public fear amplified by media and moral authorities. This process involves the identification of 'folk devils'—symbolic representations of the threat—and results in heightened social anxiety and calls for control measures, which can include new draconian laws or policies. The Panicocene extends this dynamic further by capturing how episodic social fears are turned into a prolonged, systemic crisis of collective anxiety over climate change and its attendant mass migration. The concepts of the moral panic and the Panicocene are both based on magnifying fear around perceived threats to core societal values and survival; but while the former traditionally focuses on discrete social groups or behaviors, the latter encapsulates our moral and existential crisis on a planetary scale. Thus, the Panicocene can be seen as a macrocosmic moral panic, where the 'folk devils' are not discrete social groups but rather climate change and, consequently, mass migration through the global South-to-North pipeline (Fig. 7.2).

To recast security from a logic of surveillance to one of care, we must first broaden our understanding of what being 'safe' means. Rather

Fig. 7.2 Border Birds, 2022–2025. Photographs of birds crossing borders captured with the help of open cameras and AI. This was achieved by creating software that detects birds on footage captured by open surveillance cameras placed at boundary lines. The software used artificial intelligence running on a 24/7 server between the dates of March 10th and April 10th, 2022. The server captured more than 3474 birds through multiple cameras around the world. Later, a curated selection of 100 Border Birds was made. The monitored cameras at the borders of Mexico/United States, Morocco/Spain, Greece/Turkey, and France/England. 50% of the revenue from this project goes to the European Network of Migrant Women and Red Cross supporting refugees. Courtesy of the artist Dries Depooprter (project in collaboration with Bieke Depoorter)

than measuring security in terms of barriers erected or CCTV cameras deployed, we should foreground outcomes like democracy, health, education, social cohesion, and ecological resilience. By reorienting our institutions and cultures toward solidarity and mutual aid—recognizing that each person's well-being is bound up with our own—we can move from a world policed by fear to one sustained by care.

We often forget that true security only comes with death, yet we allow ourselves to be tossed between manufactured fears. We endlessly turn to states, markets, and religions for illusory 'Security' and 'security measures.' Mark Neocleous argues that security has become the sovereign's pact against death, even as the modern state guarantees only the perpetual 'insecurity' of the capitalist order (Neocleous, 2019, p. 31). To break free from this cycle of managing terror, we must 'learn how to die'—to embrace our shared vulnerability as the foundation of our common humanity. As Neocleous suggests, practicing a genuine awareness of death—of learning how to die—may be our path to liberation from the unending jargon of insecurity (p. 32). Contrary to what one might imagine, our health as a species won't be safeguarded by a bigger-and-better border or an escape to another planet. Our survival and flourishing can only come from a new understanding of community with all living creatures on this planet. We need a communal body that is not defined in terms of the politics of identity or nationality, perhaps a parliament of all (vulnerable) bodies living on Earth.

Embracing vulnerability as the essence of human existence transforms dependence on others from a burden into the basis for connection, solidarity, and change. This perspective requires recognizing our connections not only with other humans but with the more-than-human world. As Astra Taylor (2023) argues, 'only More-Than-Human Security will suffice,' emphasizing that we must work with nature rather than against it. We must cooperate with the sun and wind for renewable energy, with the oceans and forests to sequester carbon, with biodiverse plants to cool our cities and feed populations, and with animal allies like water-protecting beavers who provide wetland refuge for other species.

In addition to working with nature, we need to collaborate with each other, which means transforming our anxiety and climate insecurity into solidarity. We need a solidarity strong enough to overcome the special interests spearheading the inadequate solutions presented in governmental climate policies. We should aspire to a more forward-looking vision of a state that provides security for all in a sustainable manner, a state that is at the same time decarbonized and democratic, what we would like to call a solidarity state. Rooted in collaboration for the commons, a solidarity state aspires to political and economic equality as well as recognition of our more-than-human interdependence.

The transformation toward care-based approaches cannot happen if it is not anchored in material conditions. As Naomi Klein (2007; 2015) has emphasized, ensuring baseline material security for people, especially through initiatives like a green jobs guarantee, is critical for facilitating a just transition away from fossil fuels. Material security can help address 'the crisis of empathy in a warming world' by reducing vulnerability to forces of racist demagoguery that prey on fears permeating times of great change.

Leading scientists echo this view, emphasizing that our future security requires attention to the interconnections between human well-being and that of animals and ecosystems. We need to invent a new community that de-links from the identity and border politics with which we have produced sovereignty until now, but that also moves beyond the reduction of life to cybernetic bio surveillance. To sustain planetary life in the face of centuries of ecological and cultural destruction, we must invent new forms of global cooperation and structures capable of coordinating care at every scale.

In reflecting on the urgent challenges of climate mobility justice, it becomes clear that the frameworks, narratives, and practices we rely on to understand and respond to these issues must themselves be transformed (and be transformative). The manifesto of *Undisciplining Political Ecology*[2] calls on us to rethink not only what we study but how we study it, inviting us to break free from the constraints of disciplinary silos and the colonial legacies embedded within academic knowledge production. Political Ecology began as a militant field dedicated to challenging systems of oppression and environmental destruction. Yet, over time, it has become at risk of being institutionalized, its radical potential diluted by the very structures it once sought to disrupt. This tension mirrors the broader challenges in climate mobility discourse, where dominant narratives often replicate colonial and securitarian logics, framing migration as a problem rather than a symptom of deeper injustices. An undisciplined approach refuses those constraints: it embraces complexity

[2] https://undisciplinedenvironments.org/2019/10/01/undisciplining-political-ecology-a-minifesto/

and vulnerability, recognizing that knowledge is not a fixed commodity but a living, relational process. It asks us to interrogate our own privileges and positionalities and the ways in which academic training shapes our understanding of the world. This self-reflection is vital if we are to move beyond abstract theorizing toward grounded, accountable, and empathetic engagements with those most affected by climate change and displacement. Moreover, undisciplining invites us to create spaces of solidarity and care within and beyond academia. It challenges the myth of objectivity and instead values the power of the arts, storytelling, poetry, and other creative forms as legitimate and necessary ways of knowing. These modes of expression resonate deeply with the principles of cosmovisional justice, which emphasizes relationality, reciprocity, care, and the sacred bonds between humans and the more-than-human world.

In this sense, the act of undisciplining becomes a political and ethical commitment: a deliberate rejection of the boundaries set by conventional disciplines and institutional norms. It is a call to cultivate new imaginaries, to embrace contradiction and uncertainty as fertile grounds for innovation, and to center long-silenced and marginalized voices. For those engaged in climate mobility justice, this means reimagining narratives that do not simply quantify risk or securitize movement but instead acknowledge historical injustices, planetary interconnectedness, and the diverse cosmovisions offering alternative ways of relating to place and mobility. It calls for a decolonial praxis that is undisciplined in its refusal to be contained, open to a full multiplicity of knowledges, and committed to collective liberation. The academic collective *Reimagining Mobilities*[3] exemplifies how the act of undisciplining can become both a political and ethical project. By refusing to accept the constraints imposed by traditional academic disciplines and institutional norms, the collective cultivates spaces for dialogue, creativity, and critical engagement. This approach is not merely about breaking boundaries for their own sake; rather, it is a conscious call to imagine new possibilities for knowledge production and dissemination. In doing so, *Reimagining Mobilities* demonstrates that undisciplining is an act of resistance—a deliberate move to foster more inclusive, pluralistic, and just forms of scholarship and social engagement.

[3] https://reimaginingmobilities.org/en/

The Panicocene, this era of pervasive panic we live in, both feeds and is fed by the moral panic directed at climate change and migration—and, consequently, at the panic (squared) evoked by climate migration (Giacomelli, 2023; Giacomelli & Cappi, 2025). To escape its grip, we must first acknowledge how every narrative—from media headlines to academic articles—shapes public perceptions of climate (im)mobilities. As earlier chapters demonstrated, the language and stories we tell about these connections have the power to steer broader representations and responses. Moreover, the narrative used to describe people on the move has the power to shape their embodied experience: whether they are treated humanely or with violence, whether they are greeted warmly or rejected with hostility. Media-driven perceptions are likely to result in public pressures that, in turn, can influence behavior and decisions, including at the policy level. Talking about the connections between climate and migration therefore carries a distinct respons-ability. All too often, contemporary media coverage downplays those responsible for the climate crisis, sidelines real solutions, and ignores the social inequalities at play, instead fixating on environmental disasters and the economic 'costs' of ecological transition. At the same time, sensationalist reporting casts migration as a monolithic South-to-North onslaught, stoking further securitized and racialized anxieties. This discursive battleground reveals deeper epistemological shifts at play: in the negative, a tension between crisis frameworks used to justify securitization measures and, in the positive, rights-based paradigms that assert the fundamental right to perform mobility with dignity.

Addressing this communication gap requires greater awareness and more accurate information that goes beyond episodic reporting of territorial destruction to promote a broader vision embracing the phenomenon in its full global scope. We need news media coverage that communicates urgency effectively while developing intersectional frameworks of climate justice that tackle climate change as a political and ethical problem, stressing both differential responsibilities and the structural inequalities underlying uneven vulnerabilities.

The challenge lies in finding the right balance between communicating with the right incisiveness while avoiding the kind of catastrophism that risks perpetuating incredulity ('this claim is too extreme') or utter demotivation ('we will never succeed'). Projects like *Framing Climate Justice*[4] offer a path forward, showing us how to actively explain the ways in which climate change multiplies oppression—as opposed to merely stating that it does—and how to pivot toward stories of solidarity and shared humanity.

Echoing this sensibility, George Monbiot (2017) suggests that 'rather than arrogating naming rights to themselves, professional ecologists should recruit poets and cognitive linguists and amateur nature lovers to help them find the words for what they cherish.'[5] This call for inter— and transdisciplinary collaboration reframes climate communication as not solely a scientific or technical challenge, but a creative, ethical, and affective practice as well. Indeed, it is often artists and activists—rather than researchers or policymakers—who most clearly recognize and act upon this need. Consider, for instance, the growing use of art and cultural institutions by climate activists to expose political inertia and demand urgent action.

This insurgent spirit is captured well by Marco Armiero's notion of 'narrative guerrilla warfare' a political gesture defined by the author as a 'stance against toxic narratives' (Armiero, 2021, pp. 22–25). His *ToxicBios*,[6] a narrative guerrilla warfare project, includes a collection of interviews on global environmental injustice that gives shape to a more extensive open-access multimedia archive. Inspired by *EJAtlas* (*Environmental Justice Atlas*), the project speaks of stories of resistance, proposing an antagonistic narrative that can potentially transform communities (Armiero, 2018). Narrative guerrilla warfare resonates conceptually with the counter-stories defined by Solórzano and Yosso (2002) 'as a method of telling the stories of those people whose experiences are not often told (i.e., those on the margins of society). The counter-story is also a tool for exposing, analyzing, and challenging the majoritarian stories of racial privilege. Counter-stories can shatter complacency, challenge the dominant discourse on race, and further the struggle for racial reform. Yet, counter-stories need not be created only as a direct response to majoritarian stories. [...] Indeed, within the histories and lives of people of color,

[4] https://framingclimatejustice.org/

[5] https://www.theguardian.com/commentisfree/2017/aug/09/forget-the-environment-new-words-lifes-wonders-language

[6] https://toxicbios.org/#/stories

there are numerous unheard counter-stories. Storytelling and counter-storytelling these experiences can help strengthen traditions of social, political, and cultural survival and resistance' (2002, p. 32).

Another proposal comes from all those communal actors seeking to reconstruct the complexity of the migratory path 'in its entirety'—that is, from its beginning—therefore connecting the 'here' and 'there' of today's wandering humanity together with the narrative resources capable of representing it. In Italy, the Archivio delle Memorie Migranti[7] and the DiMMi[8] project of the Archivio Diaristico Nazionale of Pieve S. Stefano (multimedia migrant diaries) represent just two of the organizational efforts to delicately and carefully collect and disseminate stories, narratives, and autobiographical accounts of people of foreign backgrounds. These first-person voices of migration function as repositories capable of preserving the testimonies of the new 'wretched of the earth' (Fanon, 2001) whose words, signs—and even silences—must be preserved for future generations (Triulzi, 2012).

Interrogating our imaginaries through sources that come from the margins helps us understand how perceptions of climate and mobility are constructed. The Archive of Colonial Memories,[9] for instance, contributes to the decolonization of narratives by serving as a liminal space where colonial pasts, neocolonial presents, and decolonial futures intersect. It centers indigenous memories, objects, communities, and thinkers, using these as entry points to diverse cultural contexts that challenge Eurocentric perspectives. By Africanizing the history of objects and adopting a multidisciplinary and multi perspective approach, this archive subverts hegemonic paradigms and meaningfully reclaims memories used to construct counter-narratives and identities of formerly colonized peoples. This process of 'rememory-ing' enables the repairing of social memory, forges ethical relationships to the past, and promotes a critical engagement that destabilizes colonial power structures embedded in traditional archives. Thus, the Archive of Colonial Memories acts not only as a site of remembrance but as a tool for decolonial knowledge production and social justice oriented toward newly pluralistic legacies and futures.

[7] https://www.archiviomemoriemigranti.net/

[8] https://www.archiviomemoriemigranti.net/dimmi/

[9] https://www.memoriecoloniali.org/

Alongside the examples found in Chapters 5 and 6, these proposals merely scratch the surface of what it means to build a new toolbox of words, images, and metaphors for telling the intertwined story of climate change and migration. To contrast reproductions of blind power and violence, we must renew our collective imagination through the vitality of an intersectional and decolonial approach grounded on the interdependence of care—care of the self, care of the community, and care of the Earth (Giuliani, 2021). As Achille Mbembe (2021) reminds us, 'planetary habitability' is inseparable from the politics of care, repair, and restitution. Indeed, climate justice extends far beyond environmental concerns to encompass social and civil rights, presenting economic, political, and cultural dimensions as deeply ecological ones. Above all, the fate of climate justice depends on the stories we tell—and refuse to tell—about these complex entanglements.

To transcend the Panicocene, we must cultivate new *ecotones* of imagination. In nature, ecotones are all those transition environments—natural or anthropized—interposed between two or more different biological communities, or habitats. Ecotones contain species belonging to the bordering communities and others exclusive to the ecotonal area itself, therefore possessing a rich biodiversity. This ecological peculiarity makes the ecotone indispensable because it is through these areas that the connection and exchange between very different environments (woods-meadows, lakes-forests, fresh water-salt water) takes place. Echoing Bell Hooks' praise of the margins (1989), we elevate ecotones as those in-between zones that are necessary for the flourishing alternative imaginaries through the cultivation of radical practices that are at once cultural, aesthetic, and critical. Like the notion of marginality, ecotones are spaces of radical possibility and resistance against those processes , of 'monstrification' (Giuliani, 2016) and 'brutalization' (Mbembe, 2020), wrought by hegemonic media discourses.

The Panicocene as a hegemonic discourse is present not only in words and images but also in ways of being and living. Ecotones and marginality are spaces that offer a radical perspective from which to look, create, and imagine alternatives and new worlds (hooks, 1989). They provide counter-languages to the prevailing panic that rages in the narratives of climate (im)mobilities and colonizes our minds and hearts. The lenses through which we interpret the facts of our present (and possible futures) are never clear or neutral. We need different words and tools, we need universal sisterhood and brotherhood, we need intersectional lenses: class, race, and sexuality can no longer be separated, and indeed overlap with other forms of discrimination, such as ableism and speciesism.

In the Panicocene, narrating climate (im)mobilities appropriately requires us to decenter our perspective and observe them from the margins. We must also become 'acrobats of time' (Benedetti, 2021) and put ourselves in the shoes of those who will come after us. This temporal empathy, as Judith Butler (2015) suggests, allows us to recognize ourselves as co-vulnerable across time and space. The Panicocene is characterized by mainstream discourses where the 'we' is constantly at risk of 'invasion' or 'under siege, 'so it is imperative that we expand the imaginary of our collective subjectivity to a truly intersectional, global and cosmovisional 'we.'' To date, the language and images used in the narratives on climate (im)mobilities have conveyed the idea that the environmental migrant—-affected by the climate crisis and forced to migrate—-is always the 'other,' . No longer. It is incumbent upon us to redefine that 'we' as the common destiny of humanity. Speaking of climate change, Kari Marie Norgaard (2011) cautions that the greatest catastrophe is when catastrophe itself vanishes from view—an insight tragically apparent in the world's response to the unfolding genocide in Gaza. As legal debates circle around whether the situation in Gaza meets the strict definition of genocide, the power of narrative reveals a harsher truth that legal technicalities often obscure. Tens of thousands of Palestinians have been killed, essential infrastructure destroyed, and 2 million people—over 90% of Gaza's population—displaced, with countless children dying from starvation and lack of medical care. Legal narratives demanding proof of explicit intent risk minimizing this reality and delay urgent action. The truth is that the systematic destruction and deprivation inflicted on Gaza amount to genocide in practice, if not in theory. The International Criminal Court's indictment of Netanyahu for crimes against humanity transcends legal nuance as the harrowing narratives emerging from Gaza compel the world to confront genocide as an undeniable lived experience.

Escaping the narratives of panic means embedding climate (im)mobilities within intersectional, decolonial scenarios that spark solidarity rather than paralysis. Catastrophe is always mediated by language and narration—and at the heart of our struggles against it is the 'inability to face the crisis of our culture' (Bollas, 2018, p. 37). In fact, the difficulty of creating a context of shared narratives and active listening is perhaps the biggest obstacle on the way to collecting and reformulating imaginaries on climate (im)mobilities. What we most urgently need are historical and political frameworks that not only humanize climate migrants—granting them a voice, a face, and agency—but also situate these self-narratives within an intersectional understanding that produces responsible, context-sensitive knowledge (Giuliani, 2021).

This underscores the urgent need for new kinds of stories—narratives that weave the personal into the political and illuminate the deep historical, structural, and geopolitical forces behind climate (im)mobilities. These tales must bridge the local and the global, be framed within the broader concept of climate mobility justice, and do more than just capture snapshots of the present moment. They must imagine alternative futures and invite us to envision the lives and landscapes of the generations to come. Only by projecting ourselves forward into worlds shaped by today's decisions can we begin to care more meaningfully for our present. This is not mere awareness; it is a call to respons-ability.

The metaphors we employ profoundly shape our understanding and responses to climate mobility. The dominant image of the lifeboat, as popularized by Garrett Hardin (1974) in his problematic 'Lifeboat Ethics,' positions wealthy nations as separate vessels in a sea of drowning humanity in such a way that it ends up justifying exclusion and resource hoarding. This metaphor has served as an intellectual foundation for what we now recognize as eco-fascism and 'petromasculinity,' ideologies that maintain that certain categories of humans are more deserving than others. Instead, we propose embracing alternative metaphors that better capture the reality of our interconnected fate. The image 'All in the same storm but in different boats' acknowledges differential vulnerabilities and responsibilities while preserving our shared planetary context. As Malcom Ferdinand (2022) writes, the ecological storm is just a different name for the modern hurricane that has been blowing in since at least 1492. In facing this tempest of coloniality, he urges us 'to build together not some Noah's ark or slave ship but a world-ship—a ship that welcomes a world between humans with non-humans on the bridge of justice' (2022, p. 242). We must reimagine our planetary 'world-ship' through a decolonial ecology—one that dismantles the colonial underpinnings of the Anthropocene and reopens the world to plural horizons (Chakrabarty, 2009, 2012). This is not an assault on humanism itself but on the 'sordidly racist' and profoundly misogynist strands of humanism forcefully denounced, as Ferdinand (2022) reminds us, by thinkers like Aimé Césaire, Olympe de Gouges, and Sylvia Wynter, alongside traditions that have long erased animal others and elevated Western norms as the universal standard. We do not want to abandon universality; we reject its vertical, Eurocentric form in favor of what Souleymane Bachir Diagne (2018) calls a 'truly universal universal' rooted in listening, exchange, and collective flourishing rather than domination.

In this spirit, we propose revisiting the idea of cosmopolitanism through a decolonial lens that affirms a common ground for justice struggles while embracing a pluriversal perspective. This departs from Enlightenment universalism, which has historically restricted the right to have rights (Arendt, 1966) to a narrow segment of humanity. Rights must be truly inclusive and universal; otherwise, they function as privileges for the few. To move beyond the Panicocene, we must question whether punitive, exclusionary measures truly bind communities together. How can calls for increased security coexist with the defense of fundamental human and more-than-human rights? Why are we so willing to allow fear to shape our worldviews and institutions? Reimagining insecurity not as a pretext for control, but as a catalyst for global solidarity is essential for a just, interconnected planetary future.

Thus far, our language and images of climate (im)mobilities have conveyed the idea that it is the 'other' who is affected by the climate crisis and forced to migrate. We must forge a new 'we,' a shared destiny that acknowledges how climate change affects us all in dramatically unequal ways. Only by situating our responses in this shared vulnerability—rather than in further illusions of separation or immunity—can we craft lasting solidarity.

The climate crisis also demands an ontological reworking of our relationship with the world and our way of 'being' in it. Haraway's (2016) injunction to 'Stay with the Trouble,' reminds us that 'no species, not even our presumptuous one that claims to be good individuals in so-called modern Western scripts, acts alone; assemblages of organic species and abiotic actors make history, both evolutionary and other kinds of history' (2016, p.100).

As we conclude this exploration of the narratives and imaginaries of climate (im)mobilities, we return to Mastrojeni & Pasini's (2020) call to 'transform uncertainty into an opportunity.' The climate crisis and its impacts on human mobility present profound challenges to our systems, institutions, and ways of thinking. Yet they also offer an unprecedented opportunity to reimagine our relationships with each other and with the Earth's systems that sustain us.

By centering the experiences and agency of those most affected, acknowledging differential responsibilities, privileging an ethics of care over an ethics of insecurity, and crafting metaphors that better capture our interdependence, we can begin to build more just and sustainable responses to climate mobility. This approach moves beyond the binary of adaptation versus mitigation, recognizing that how we narrate and respond to climate-induced displacement is itself a form of climate action with profound implications for justice.

At this defining moment, the reflections, narratives, and representations gathered here insist that socio-political issues must stand alongside concern for the climate crisis and that holding together this linkage should be at the core of how we understand and respond to climate mobilities. By grounding our responses in care—for self, community, and the Earth—and by holding out against reproductions of power and violence, we lay the groundwork for bridging theory and policy and fighting for empirically grounded, equality-oriented, and actor-driven solutions to one of the most important challenges of our generation.

The right to breathe (Mbembe, 2020), to move, to stay in place, and to thrive are not privileges to be granted or denied; they are fundamental to belonging in the biosphere and participating in its life. Acknowledging our shared planetary stake—and working together across differences—transforms our response to climate (im)mobilities from a politics of fear and exclusion to one of care and solidarity. This transformation is not optional but necessary for creating a future in which both human and more-than-human communities can flourish in the face of climate change.

Ultimately, the struggle for climate mobility justice is a struggle over meaning, representation, and power. It is about who gets to tell the story of climate change, whose futures are imagined, and whose lives are valued. By embracing planetary politics, cosmovisional justice, decolonial approaches, and undisciplined ecologies, we can begin to dismantle the coloniality of climate narratives and breathe life into new forms of solidarity, responsibility, and planetary care.

References

Arendt, H. (1966). *The origins of totalitarianism*. Penguin Classics.

Armiero, M. (2018). Epichange. Toxic bios. Autobiografie tossiche: Un progetto di guerriglia narrativa. *Epidemiologia & Prevenzione, 42*(3–4), 212–213. https://doi.org/10.19191/EP18.3-4.P212.068

Armiero, M. (2021). *Wasteocene: Stories from the global dump (1st ed)*. Cambridge University Press.
Benedetti, C. (2021). *La letteratura ci salverà dall'estinzione*. Giulio Einaudi editore.
Boas, I. (2025). *Beyond climate migration: The politics and practice of climate mobilities*. doi:https://doi.org/10.18174/687963.
Boas, I., Farbotko, C., & Bukari, K. N. (2024). The bordering and rebordering of climate mobilities: Towards a plurality of relations. *Mobilities, 19*(3), 521–536. https://doi.org/10.1080/17450101.2023.2279095
Bollas, C. (2018). *L'età dello smarrimento: Senso e malinconia*. Cortina.
Boltanski, L. (1999). *Distant suffering: Morality, media, and politics*. Cambridge University Press.
Boyd, A. (2024). Switzerland and Italy redraw border due to melting glaciers. *BBC, 20 September*. https://www.bbc.com/news/articles/cgk7r0rrdnmo
Butler, J. (2015). 2. Bodies in Alliance and the politics of the street. In *Notes toward a performative theory of assembly* (pp. 66–98). Harvard University Press. https://doi.org/10.4159/9780674495548-003
Castel, R. (2011). *L'insicurezza sociale. Che significa essere protetti?* Einaudi.
Chakrabarty, D. (2009). The climate of history: Four theses. *Critical Inquiry, 35*(2), 197–222. https://doi.org/10.1086/596640
Chakrabarty, D. (2012). Postcolonial studies and the challenge of climate change. *New Literary History, 43*(1), 1–18.
Chouliaraki, L. (2013). *The ironic spectator: Solidarity in the age of post-humanitarianism*. Polity Press.
Cohen, S. (2011). *Folk devils and moral panics: The creation of the mods and rockers*. Routledge.
Diagne, S. B. (2018). Faire humanité ensemble et ensemble habiter la terre: *Présence Africaine. N°, 193*(1), 11–19. https://doi.org/10.3917/presa.193.0011
Draper, J. (2024). *Climate Displacement*. Oxford University Press.
Fanon, F. (2001). *The wretched of the earth (reprinted)*. Penguin Books.
Farbotko, C. (2022). The specter of mass climate migration across international borders: Dismantling an unscientific expectation. *One Earth, 5*(8), 841–844. https://doi.org/10.1016/j.oneear.2022.07.009
Ferdinand, M. (2022). *Decolonial ecology: Thinking from the Caribbean world* (A. P. Smith, Trans.). polity.
Ferreira da Silva, D. (2018). *On Heat*. https://canadianart.ca/features/on-heat/
Giacomelli, E. (2023). *Panicocene: Narrazioni su cambiamenti climatici, regimi di mobilità e migrazioni ambientali*. Franco Angeli.
Giacomelli, E., & Cappi, V. (2025). 'Othering' in the Panicocene: Intergovernmental and non-governmental awareness campaigns on climate change

induced-migration. *Journal of Ethnic and Migration Studies, 1–19*. https://doi.org/10.1080/1369183X.2025.2468287

Girard, R. (1972). *La violence et le sacré* (Nouv. tir). Grasset.

Giuliani, G. (2016). *Zombie, alieni e mutanti: Le paure dall'11 settembre a oggi* (Prima edizione). Le Monnier.

Giuliani, G. (2021). *Monsters, catastrophes and the anthropocene: A postcolonial critique*. Routledge.

Haraway, D. J. (2016). Staying with the trouble: Making kin in the Chthulucene (p. dup; 9780822373780/1). Duke University Press. doi:https://doi.org/10.1215/9780822373780.

Hardin, G. (1974). *Lifeboat ethics: The case against helping the poor by Garrett Hardin*. Psychology Today.

Harvey, S. (2023). *Orbital (first grove Atlantic hardcover edition)*. Grove Press.

Hooks, B. (1989). CHOOSING the margin as a space of radical openness. *Framework: The Journal of Cinema and Media, 36*, 15–23.

Klein, N. (2007). *The shock doctrine: The rise of disaster capitalism*. Allen Lane.

Klein, N. (2015). *This changes everything: Capitalism vs. the climate*. Penguin books.

Krenak, A. (2024). *Ancestral future*. Polity.

Lear, J. (2006). *Radical Hope*. Harvard University Press. https://doi.org/10.2307/j.ctvjz82p7

Mastrojeni, G., & Pasini, A. (2020). *Effetto serra, effetto guerra: [Il clima impazzito, le ondate migratorie, i conflitti; il riscaldamento globale, i ricchi, i poveri]* (Nuova ed). Chiarelettere.

Mbembe, A. (2020). *Brutalisme*. La Découverte.

Mbembe, A. (2021). The universal right to breathe. *Critical Inquiry, 47*(2), 58–62.

Mignolo, W. D. (2000). *Local histories/global designs: Coloniality, subaltern knowledges, and border thinking*. Princeton University Press.

Monbiot, G. (2017). Forget the «environment»: We need new words to convey life's wonder. https://www.theguardian.com/commentisfree/2017/aug/09/forget-the-environment-new-words-lifes-wonders-language

Neocleous, M. (2019). Securitati perpetuae: Death, fear and the history of insecurity. *Radical Philosophy, 206*, 19–33. https://www.radicalphilosophy.com/article/securitati-perpetuae

Norgaard, K. M. (2011). *Living in denial: Climate change, emotions, and everyday life*. MIT Press.

Nussbaum, M. C. (2018). *The monarchy of fear: A philosopher looks at our political crisis (first Simon&Schuster hardcover edition)*. Simon & Schuster.

Papastergiadis, N. (2023). *The cosmos in cosmopolitanism*. Polity Press.

Quijano, A. (2007). COLONIALITY and modernity/rationality. *Cultural Studies, 21*(2–3), 168–178. https://doi.org/10.1080/09502380601164353

Robin, C. (2004). *Fear: The history of a political idea*. Oxford University Press.
Boaventura de Santos, S. (Ed.). (2010). *Voices of the world*. Verso.
Sheller, M. (2018). *Mobility justice: The politics of movement in the age of extremes*. Verso.
Solórzano, D. G., & Yosso, T. J. (2002). Critical race methodology: Counter-storytelling as an analytical framework for education research. *Qualitative Inquiry, 8*(1), 23–44. https://doi.org/10.1177/107780040200800103
Taylor, A. (2023). *The age of insecurity: Coming together as things fall apart*. House of Anansi Press Inc.
Triulzi, A. (2012). Per un archivio delle memorie migranti. *Zapruder, 28*, 120–125.
Yumagulova, L., Parsons, M., Yellow Old Woman-Munro, D., Dicken, E., Lambert, S., Vergustina, N., Scott, J. C., Michell, P., & Black, W. (2023). Indigenous perspectives on climate mobility justice and displacement-mobility-immobility continuum. *Climate and Development, 1–18*. https://doi.org/10.1080/17565529.2023.2227158

Afterword

Professor Andrew Baldwin, Durham University

If I am honest, thoughts of climate change have never really worried me that much. Nor has the thought that the impacts of climate change will displace millions of people around the world. What really worries me though is the fragile nature of political life. That at a moment's notice, under the veil of benevolence, contemporary political life in the west can give way in an instant to something rather sinister. More than future climate change, I worry about what is happening in Gaza, in Ukraine and in America. I worry about the society of control forming in the ideological fog of Silicon Valley. I worry about the rising popularity of authoritarianism in Europe. And I worry that racisms find their clearest expression in their vehement denial but also form the unconscious substrate of desire itself.

It is not that I do not care about climate change. In fact, I have spent much of my academic life thinking and writing about the relationship between climate change and human migration. But I have done so for reasons less to do with climate change per se than out of a concern that the darker side of human political life will materialise as climate change becomes a reality. I have no reason to think that those who benefit from racist structures of domination will not double down on those very structures as a strategy for ensuring their survival as planetary

E. Giacomelli and P. Musarò, *Climate Mobility Justice*,
https://doi.org/10.1007/978-3-032-07907-7

habitability declines. In fact, my primary worry with climate change is that existing structures of domination may start to adapt in response to climate change in less than obvious ways. My early readings of Foucault got me thinking that if planetary liveability was called into question, the unfortunate among us may well end up cast from the realm of the living precisely so that others may go on living through the planetary crisis.

Eventually I would start to examine how climate change and racism were newly converging around questions of migration and displacement, a point Giacomelli and Musarò make all too clear in *Climate Mobility Justice*. And this led me to the insight that if the goal is climate justice, whatever that might mean, then surly undoing racism, which is a structural feature of contemporary political life across the West, would need to be part of the mix. This seems straightforward enough. But this is also where my thinking starts to get a bit more complicated. I developed a nagging suspicion that while calls for reparations, like that put forward by Olúfémi Tàìwó (Táíwò, 2022), are important for bringing historical accountability into discussions about climate change, a new form of racism was also forming in the shadow of climate change discourse largely unnoticed. My concern was that without taking account of this new racism, a new spatial structure of domination would start to emerge akin to what some are nowadays calling 'climate apartheid' (Long, 2024). And so while climate change itself does not keep me up at night, the political unconscious of climate chance discourse does. I worry that without a concise theory of racism, responses to climate change on the part of national governments, international institutions and markets may end up doing more harm than good. Deciphering this new racism has been a long-standing preoccupation of mine.

All of this is a very long-winded way of saying that *Climate Mobility Justice* is important in my view because it helps us better appreciate how race and racism work at the intersection of climate change and mobility. Giacomelli and Musarò's propose, for example, that we shift our 'analytical attention away from static territorial units (nation-states, societies) or isolated actors (individuals, groups, communities) to the power structures that regulate mobility: from border architectures to transportation infrastructures and the multi-scalar governance systems that structure circulation and containment.' What their analysis points toward is that racism—a structure of domination that is as much discursive as material—is not reducible to the actions of a state or individual, although it *can be* both, but functions a means of mediating the world. That is to say, racism is a system of meaning. This is why Giacomelli and Musarò place so much emphasis, first and foremost, on narrative and aesthetic. Narrative and aesthetics are central to how our lives are made meaningful and they do this by shaping, sometimes explicitly and sometimes implicitly, how certain people are understood to belong more than others. Among the core observations Giacomelli and Musarò make in their book is that we can read these systems of belonging through narrative as much as through socio-technical systems like transportation infrastructure—one of the core conceits of the mobilities paradigm. Ultimately, what *Climate Mobility Justice* gives us then is an entry point for conceptualizing the dialectical relation between belonging and mobility as it pertains to the planetary crises of climate change and the Anthropocene. This is a broad intellectual remit but also an urgent one. Their demand is that we give this confluence of ideas its due, both analytically and politically.

But another reason *Climate Mobility Justice* is important is that it maps out the contours of an expansive political vision—or rather *cosmovision*—that can move climate justice discourse beyond simply a concern with displacement and human rights, important though this is, to a more encompassing field of relations in which human and non-human survival on a planet in ruin rests on a kind of universal mobility. This is a positive, affirmative cosmopolitical vision that is simultaneously decolonial and hopeful. It weaves together Achille Mbembe's notion of planetarity, which foregrounds the universal right to breathe as an essential attribute of political solidarity, with Bruno Latour's parliament of things, which fosters a universality of belonging in which human and non-human entities are permitted to coexist in the quest to bring about more liveable futures on a planet whose habitability cannot be completely guaranteed. In doing so, *Climate Mobility Justice* provides something of a model for Ian Baucom's (Baucom, 2020) heterochronic philosophy of freedom for the Anthropocene. Baucom sets out his dialectical philosophy in his book *History 4°C: Search for a Method for the Anthropocene*. In it he combines a reading of the material forces of history e.g., exploitation, extraction, violence, with the material forcings of the planet, including non-human entities, and squares these very different materialisms with the planetary humanisms of Paul Gilroy and Mbembe. Baucom's book and *Climate Mobility Justice* complement one another well. Reading them together could prove fruitful for deepening the approach that Giacomelli and Musarò develop.

It strikes me too that it is precisely this kind of cosmovision that is necessary for reconceptualising the relationship between climate change and human mobility. For too long the debate about this relationship has been focused on how best to govern human mobility (or immobility) in the context of climate change. The operating assumption is that certain forms of mobility like migration and displacement are inherently problematic and thus warrant some form of governmental intervention. Such intervention is often imagined in terms of risk management, a pre-emptive form of governance that seeks to prevent chaos and disorder from arising. But the cosmovision that Giacomelli and Musarò sketch out in their book is different. Whereas risk management is often based on a catastrophic imaginary, these authors set out a positive vision that invites consideration of what mobility in the context of climate change could look like. Rather than ask how to manage a growing crisis, they provide us with a vocabulary that inspires a more positive political imaginary. A future that all can to aspire to rather than one to be avoided. And here's the political rub. Up until recently so much of the critique about climate change and migration has taken aim at the logic of climate reductionism, which, by reducing the future to climate change, depoliticizes the future as a political resource. The logic of climate reductionism says that the future is known in advance and that it is the task of government to either prevent it from materializing or prepare populations for its eventualities. The cosmovision set out by Giacomelli and Musarò, by contrast, reclaims the future as a political resource. It approaches the future as an open horizon that inspires rather than deadens the political imaginary.

And this brings us to why I think *Climate Mobility Justice* is worthwhile. Climate change does not really worry me. What worries me is a future devoid of hope and popular democracy. Giacomelli and Musarò's cosmovision is an optimistic form of political thought that can guide us in the task of democratic renewal for a liveable future.

References

Baucom, I. (2020). *History 4° celsius: Search for a method in the age of anthropocene*. Duke University Press.

Long, J. (2024, June). Reckoning climate apartheid. *Political Geography, 112*, 103,117.

Táíwò, O. O. (2022). *Reconsidering reparations: Worldmaking in the climate crisis*. Oxford University Press.

Index

E. Giacomelli and P. Musarò, *Climate Mobility Justice*,
https://doi.org/10.1007/978-3-032-07907-7

D

FSC
www.fsc.org
MIX
Papier | Fördert gute Waldnutzung
FSC® C083411

Kolibri 360 GmbH
Ferdinand-Jühlke-Straße 7
99095 Erfurt, Deutschland
produktsicherheit@kolibri360.de